Contents

SECTION 1 Changing river environments
- 1.1 The main hydrological characteristics and processes which operate in rivers and drainage basins — 4
- 1.2 The main landforms associated with these processes — 7
- 1.3 Rivers present opportunities and hazards for people — 9

SECTION 2 Changing coastal environments
- 2.1 Physical processes that shape the coast — 12
- 2.2 The main landforms associated with these processes — 14
- 2.3 Coasts present opportunities and hazards for people — 16

SECTION 3 Changing ecosystems
- 3.1 Characteristics of the Antarctic ecosystem — 21
- 3.2 Threats to the Antarctic ecosystem and how they can be managed — 24
- 3.3 Characteristics of the tropical rainforest ecosystem — 27
- 3.4 Threats to the tropical rainforest and how they can be managed — 29

SECTION 4 Tectonic hazards
- 4.1 The structure of the Earth and distribution of earthquakes and volcanoes — 32
- 4.2 The processes and features associated with earthquakes and volcanoes — 34
- 4.3 The impacts of tectonic hazards — 37
- 4.4 Managing the impacts of tectonic hazards — 40

SECTION 5 Climate change
- 5.1 The natural and human causes of climate change — 43
- 5.2 The impacts of climate change at a range of geographical scales — 45
- 5.3 Responses to climate change — 49

SECTION 6 Changing populations
- 6.1 Populations grow and decline — 51
- 6.2 Population structures change over time — 54
- 6.3 The causes and impacts of international migration — 56

SECTION 7 Changing towns and cities
- 7.1 Where people live — 60
- 7.2 The opportunities and challenges of urbanisation — 62
- 7.3 The management of urban growth — 66

SECTION 8 Development
- 8.1 Measuring development — 68
- 8.2 The world is developing unevenly — 75
- 8.3 Achieving sustainable development — 77

SECTION 9 Changing economies
- 9.1 Changing employment structures — 82
- 9.2 The impact of globalisation and the role of transnational corporations — 86
- 9.3 Tourism is a growing industry — 88

SECTION 10 Resource provision
- 10.1 How food is produced — 90
- 10.2 Global patterns of food supply and demand — 91
- 10.3 The challenges of food supply — 93
- 10.4 How our energy is produced — 96
- 10.5 The global patterns of energy supply and demand — 98
- 10.6 The impacts of energy production — 100

Geographical skills — 102

1 Changing river environments

1.1 The main hydrological characteristics and processes which operate in rivers and drainage basins

1 Define the following terms:

   a the long profile

   ....

   ....

   b width, depth, and speed of flow/velocity

   ....

   ....

   ....

   ....

   c discharge

   ....

   ....

   d wetted perimeter

   ....

   ....

   e channel

   ....

   ....

   f watershed

   ....

   ....

   g tributary

   ....

   ....

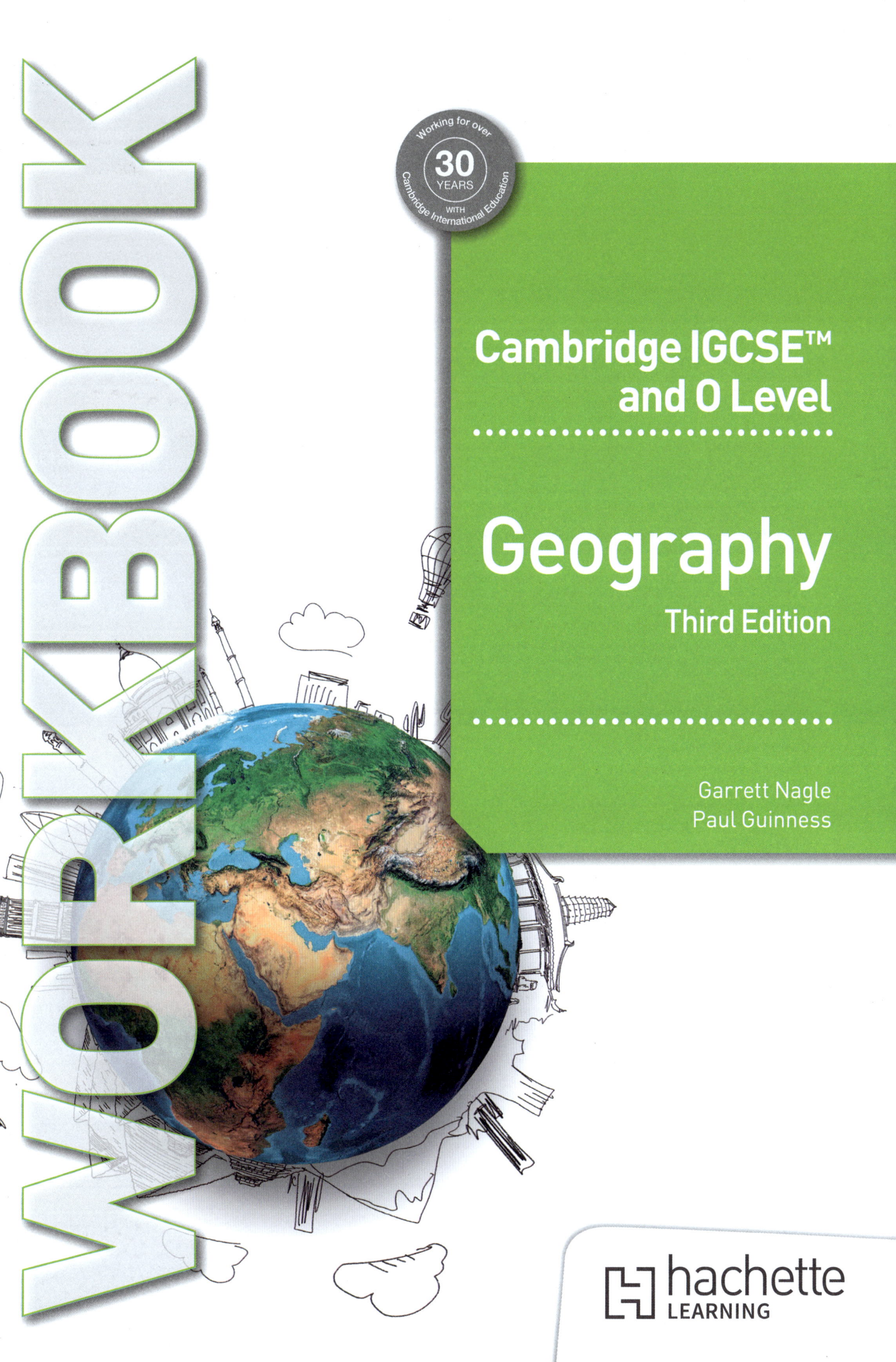

The Publishers would like to thank the following for permission to reproduce copyright material.

Photo credits
Garrett Nagle p.7, p.59, p.62; p.14 *tl* © Bennymarty/stock.adobe.com; *tr* Republic of South Africa/Department of Land Affairs/Surveys and Mapping; **p.16** www.theAA.com/travel, © KOMPASS-Karten Gmbh; **p.24** Türkiye Today, 2024 (Photo via metdesk); **p.54** ResearchGate: With permission from © Watanabe, S. et al.(2018). Longevity and elderly care: lessons from Japan. *Global Health Journal*. 10.1016/S2414-6447(19)30177-0. **p.63** © Go My Media/Shutterstock.com; **p.64** © Guy Vital-Herne / Alamy Stock Photo; **p.68** Data compiled from multiple sources by World Bank (2025) - with minor processing by Our World in Data - https://ourworldindata.org/grapher/gross-national-income-per-capita-worldbank/Creative Commons BY license; **p.75** World Bank (2024) - with major processing by Our World in Data - https://ourworldindata.org/grapher/world-bank-income-groups/Creative Commons BY license; **p.98** © Energy Institute - Statistical Review of World Energy (2024); Smil (2017) - Our World in Data/Creative Commons BY license; **p.99** BP Statistical Review of World Energy 2018, p.10; **p.106** *t and b* © BP Statistical Review of World Energy 2021 - Our World in Data/Creative Commons BY license; **p.107** © Our World in Data/Creative Commons BY license.

Every effort has been made to trace all copyright holders, but if any have been inadvertently overlooked, the Publishers will be pleased to make the necessary arrangements at the first opportunity.

Although every effort has been made to ensure that website addresses are correct at time of going to press, Hachette Learning cannot be held responsible for the content of any website mentioned in this book. It is sometimes possible to find a relocated web page by typing in the address of the home page for a website in the URL window of your browser.

Hachette UK's policy is to use papers that are natural, renewable and recyclable products and made from wood grown in well-managed forests and other controlled sources. The logging and manufacturing processes are expected to conform to the environmental regulations of the country of origin.

To order, please visit www.HachetteLearning.com or contact Customer Service at education@hachette.co.uk / +44 (0)1235 827827.

ISBN: 978 1 0360 1084 3

© Garrett Nagle and Paul Guinness 2025

First published in 2015
This edition published in 2025 by Hachette Learning,
An Hachette UK Company
Carmelite House
50 Victoria Embankment
London EC4Y 0DZ
www.hachettelearning.com

The authorised representative in the EEA is Hachette Ireland, 8 Castlecourt Centre, Dublin 15, D15 XTP3, Ireland (email: info@hbgi.ie)

Impression number   10 9 8 7 6 5 4 3 2 1

Year   2029 2028 2027 2026 2025

All rights reserved. Apart from any use permitted under UK copyright law, no part of this publication may be reproduced or transmitted in any form or by any means, electronic or mechanical, including photocopying and recording, or held within any information storage and retrieval system, without permission in writing from the publisher or under licence from the Copyright Licensing Agency Limited. Further details of such licences (for reprographic reproduction) may be obtained from the Copyright Licensing Agency Limited, www.cla.co.uk

Cover photo © Sergey Nivens – stock.adobe.com

Illustrations by Aptara, Inc.

Typeset in India by Aptara, Inc.

Printed in Great Britain by Bell & Bain Ltd, Glasgow

A catalogue record for this title is available from the British Library.

1.1 The main hydrological characteristics and processes which operate in rivers and drainage basins

h confluence

....

....

i source

....

....

j mouth

....

....

2 Describe the main features of the Bradshaw model.

....

....

....

....

....

....

....

3 Using an annotated diagram, explain how the drainage basin operates within the water cycle. Your diagram should include processes such as precipitation, interception, infiltration, percolation, overland flow, channel flow, throughflow, groundwater flow, transpiration, evaporation and evapotranspiration.

1 CHANGING RIVER ENVIRONMENTS

4 a Outline the processes of erosion, transport and deposition which operate within a river. You should include for erosion – hydraulic action, abrasion, attrition and solution; for transport – traction, suspension, saltation and solution.

....

b Define deposition and state the conditions that lead to an increase in deposition.

....

1.2 The main landforms associated with these processes

The photograph below shows a river landscape in Eastern Europe. The white areas away from the river channel are covered in snow. The sketch shows the same area.

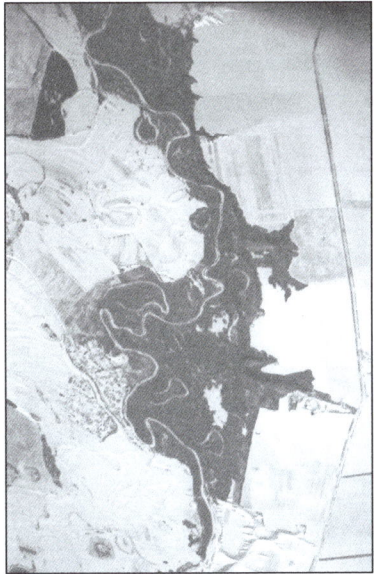 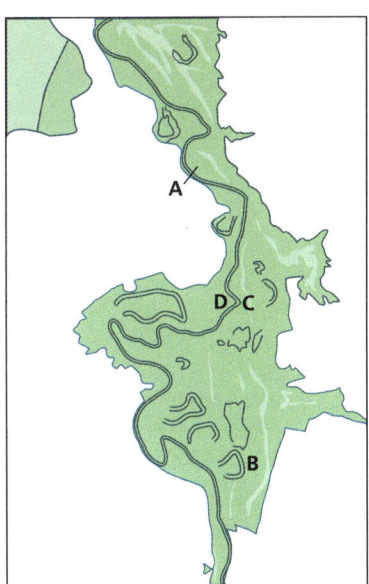

1 a Identify the landforms A and B.

A .... B ....

b Draw a cross-section to show the main features of the river channel at the meander, CD.

1 CHANGING RIVER ENVIRONMENTS

2 a With the use of a diagram, explain how an oxbow lake is formed.

b Explain how floodplains are formed.

....

....

....

....

....

....

c Explain how the snow shown in the photograph may affect the seasonal flow of water (regime) of the rivers.

....

....

....

....

1.3 Rivers present opportunities and hazards for people

1 Define the term 'flood'.

....

....

2 a Outline the disadvantages of floods.

....

....

....

....

b Comment on the advantages that floods bring.

....

....

....

....

....

3 a Identify the natural causes of floods.

....

....

....

b Outline the human factors that contribute to floods.

....

....

....

....

4 a Describe the relationship between flood magnitude and flood frequency.

....

....

....

....

1 CHANGING RIVER ENVIRONMENTS

b Explain why some floods have greater impacts than others.

....

....

....

....

5 Evaluate the strategies and techniques used to manage river pollution, including sustainable strategies for river pollution.

....

....

....

....

....

....

....

....

6 Examine the causes and impacts of a flood for a named river that you have studied. Evaluate the strategies and techniques, including sustainable strategies, used to manage the flooding of a river you have studied.

....

....

....

....

....

....

....

7 a Examine the causes and impacts of pollution in a named river.

....

....

....

1.3 Rivers present opportunities and hazards for people

....
....
....
....

b To what extent are the strategies and techniques used to manage pollution levels in the river that you have studied sustainable?

....
....
....
....
....
....
....
....

2 Changing coastal environments

2.1 Physical processes that shape the coast

1 Define the following types of coastal erosion:

   a hydraulic action

   ....

   ....

   b corrosion

   ....

   ....

   c corrasion

   ....

   ....

   d attrition

   ....

   ....

2 a Explain different types of transportation in coasts.

   ....

   ....

   ....

   ....

   b Explain why deposition occurs.

   ....

   ....

   ....

   ....

2.1 Physical processes that shape the coast

3 Distinguish between constructive and destructive waves.

   ....
   ....
   ....
   ....
   ....
   ....
   ....
   ....

4 Draw a labelled diagram to show how and why longshore drift occurs.

2 CHANGING COASTAL ENVIRONMENTS

2.2 The main landforms associated with these processes

1 Study figures below, which show a photograph and map of the Cape Peninsula in South Africa.

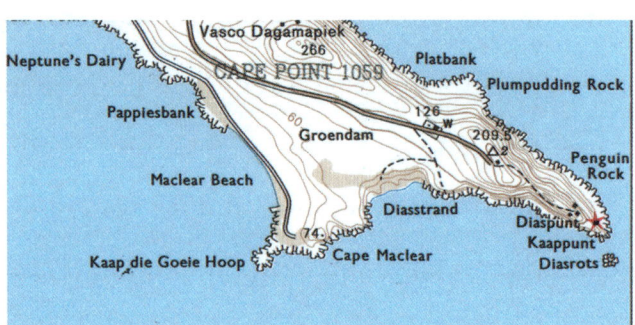

a Identify the highest point on the map and state its height.

....

....

b Compare the relief of the slopes on the eastern side (Penguin Rock) with that on the western side (Maclear Beach).

....

....

....

....

2 a Identify the landforms A, B and C on the diagram below.

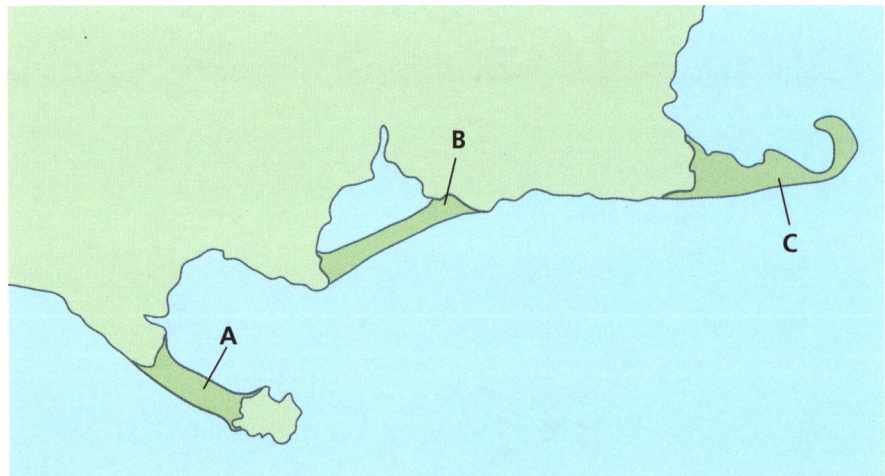

A ....

B ....

C ....

2.2 The main landforms associated with these processes

b Describe each of the three landforms.

A

....
....
....
....
....
....

B

....
....
....
....
....
....

C

....
....
....
....
....
....

c Outline the conditions needed for the formation of landform C.

....
....
....
....
....
....

2 CHANGING COASTAL ENVIRONMENTS

2.3 Coasts present opportunities and hazards for people

Study the map of south-west Tenerife and the key below.

Key

- ≋ Swimming pool
- ≋ Bathing beach
- ⌑ Diving
- ⚓ Haven, ship landing
- ☀ Lighthouse

2.3 Coasts present opportunities and hazards for people

1 Outline some of the opportunities associated with living near the coast.

   ....
   ....
   ....
   ....
   ....
   ....
   ....
   ....
   ....
   ....
   ....
   ....
   ....
   ....

2 Outline some of the hazards associated with living near the coast.

   ....
   ....
   ....
   ....
   ....
   ....
   ....
   ....
   ....
   ....
   ....
   ....
   ....
   ....

2 CHANGING COASTAL ENVIRONMENTS

3 To what extent are hard and soft engineering strategies and techniques used to manage coastal erosion and flooding sustainable?

　...

　...

　...

　...

　...

　...

　...

4 Describe the distribution and impacts of tropical storms.

　...

　...

　...

　...

　...

　...

　...

5 To what extent are preparation, planning, protection and prediction used to manage the impacts of tropical storms sustainable?

　...

　...

　...

　...

　...

　...

　...

6 Describe the global distribution of coral reefs and mangroves.

　...

　...

　...

　...

　...

2.3 Coasts present opportunities and hazards for people

7 Outline the importance of coral reefs and mangroves.

....

....

....

....

....

....

8 Analyse the threats to coral reefs and mangroves.

....

....

....

....

....

....

9 To what extent are strategies and techniques used to protect and manage coral reefs and mangroves sustainable?

....

....

....

....

....

....

10 Using a detailed, specific example of a named country or coastal area, outline the causes and impacts of coastal erosion.

....

....

....

....

....

....

2 CHANGING COASTAL ENVIRONMENTS

11 Using a detailed, specific example of a named country or coastal area, assess the sustainability of the strategies and techniques used to protect the coast from tropical storms and erosion.

....
....
....
....
....
....
....
....

3 Changing ecosystems

3.1 Characteristics of the Antarctic ecosystem

The data below shows climate data for the American Amundsen-Scott station at the South Pole and for Manaus in the Amazon.

	Jan	Feb	Mar	Apr	May	Jun	Jul	Aug	Sep	Oct	Nov	Dec	Average
South Pole (°C)	−28.2	−40.9	−54	−57.3	−57	−58	−59.7	−60	−54.9	−51.1	−38.3	−25.8	
Manaus, Amazon (°C)	28	28	28	27	28	28	28	29	29	29	29	28	

(Source: South Pole, Cool Antarctica.com; Manaus, *Philip's Modern School Atlas*)

1 a State the meaning of the term 'annual temperature range'.

....

....

b Calculate the annual temperature range for the American Amundsen-Scott station at the South Pole and for Manaus in the Amazon.

....

....

c Calculate the average annual temperature for the South Pole and Manaus.

....

....

2 a On the graphs below, complete the plot of mean average monthly temperature for the American Amundsen-Scott station at the South Pole and for Manaus in the Amazon.

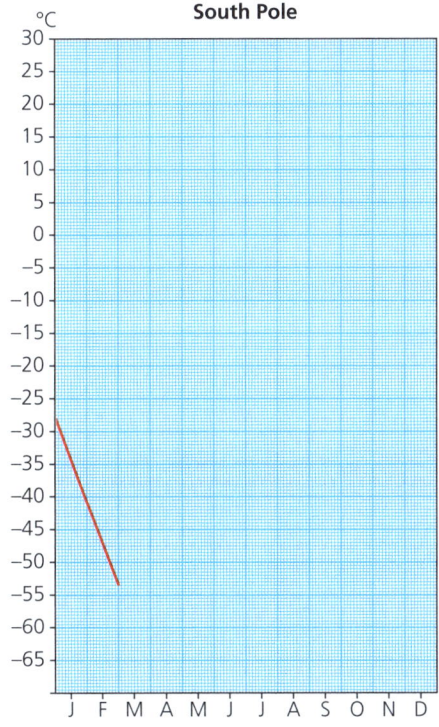
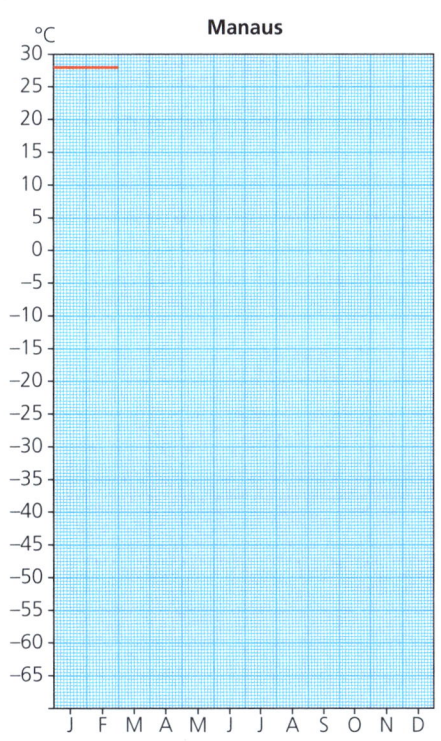

3 CHANGING ECOSYSTEMS

b Contrast the variations in the temperatures for the two locations.

....

....

....

....

....

....

The diagram below shows a food web in Antarctica.

3 State the type of environment in which this ecosystem is found.

....

4 State the source of energy in this environment.

....

5 Distinguish between food webs and food chains.

....

....

....

....

3.1 Characteristics of the Antarctic ecosystem

6 In the diagram above, identify

  a a producer

  ....

  b a herbivore

  ....

  c a primary predator/carnivore

  ....

  d a secondary predator/carnivore

  ....

  e the top carnivore in the ecosystem

  ....

7 Using examples, outline the relationship between abiotic and biotic factors.

  ....

  ....

  ....

  ....

  ....

  ....

3 CHANGING ECOSYSTEMS

3.2 Threats to the Antarctic ecosystem and how they can be managed

In July 2024, ice sheets in Antarctica experienced many temperature anomalies, such as temperatures that were significantly different from what would be expected for that time of year. Ground temperatures were, on average, 10°C above normal for July, and on some days they were as much as 28°C higher than expected. This happened as the world experienced 12 months of average temperatures of 1.5°C higher than pre-industrial level. Climate models have long suggested that the most significant impacts of global climate change would be in high latitude areas including Antarctica. Such warming could lead to the collapse of ice sheets. The heatwave was the second to hit Antarctica in two years. In March 2022, a spike of 39°C caused an ice sheet the size of Rome to collapse. Having less sea ice in the Southern Ocean and warmer temperatures in the ocean could produce warmer winters over Antarctica. Heatwaves over Antarctica are becoming more frequent.

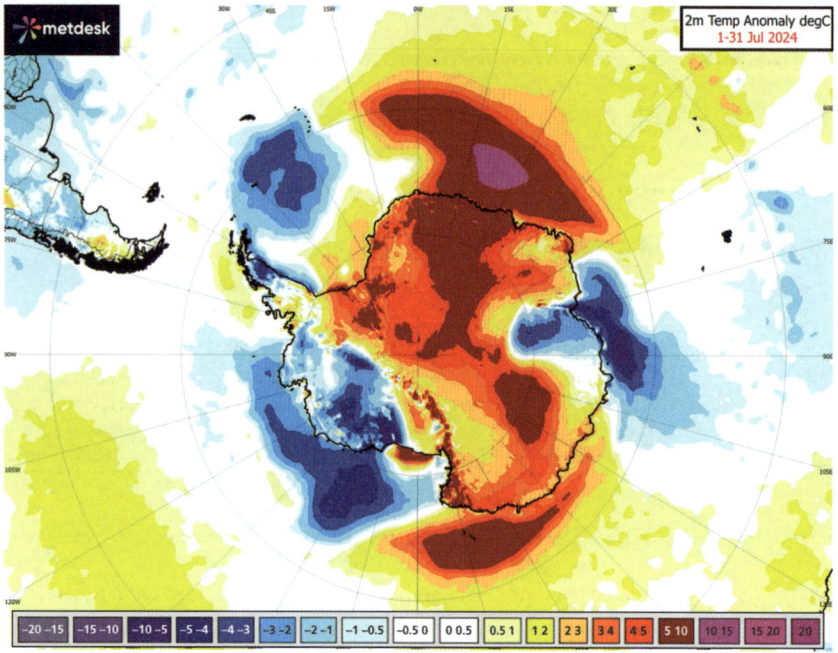

(Source: *Türkiye Today*, 2024)

1 a Using an atlas, or other resource, describe the areas that experienced a reduction in temperature in Antarctica of −2.1°C or more in July 2024.

....

....

....

....

b Using an atlas, or other resource, describe the areas that experienced an increase in temperature in Antarctica of 4°C or more in July 2024.

....

....

....

....

3.2 Threats to the Antarctic ecosystem and how they can be managed

2 a Explain the term 'temperature anomaly'.

   ....

   ....

   b State how much higher average 2m temperatures were in July 2024 than the average for Antarctica.

   ....

   ....

   c State the maximum temperature anomaly that occurred in Antarctica in July 2024.

   ....

   ....

3 Identify **two** impacts of increased temperatures on Antarctica's environment.

   ....

   ....

   ....

4 Suggest the potential link between global climate change and increased temperatures in Antarctica.

   ....

   ....

   ....

   ....

5 Name the countries that claim a right to Antarctica.

   ....

   ....

   ....

6 According to the Antarctic Treaty, how should Antarctica be used?

   ....

   ....

7 Outline the environmental protection offered by the Madrid Protocol.

   ....

   ....

   ....

3 CHANGING ECOSYSTEMS

8 According to the Convention for the Conservation of Antarctic Marine Living Resources, how should fishing be conducted?

....

....

....

9 How is whaling regulated in Antarctica?

....

....

....

....

....

10 Outline the objectives of the UN Convention on the Law of the Sea.

....

....

....

....

....

....

....

....

3.3 Characteristics of the tropical rainforest ecosystem

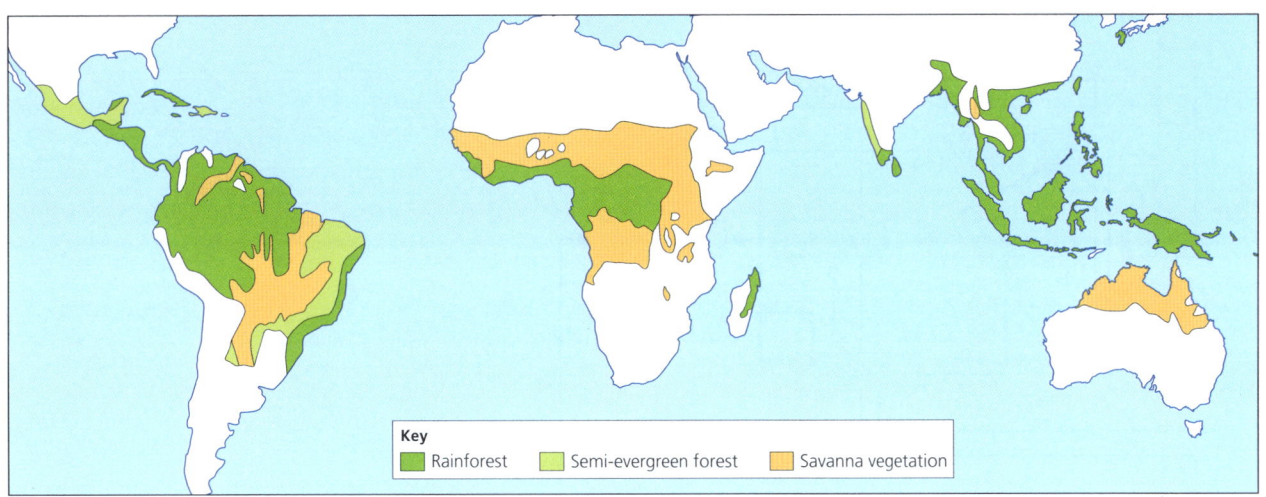

1 Describe the location of areas of tropical rainforest.

   ....

   ....

   ....

2 Identify the type of pressure system that is associated with areas of tropical rainforest.

   ....

   ....

   ....

3 State the likely climatic conditions of this environment.

   a Temperature

   ....

   ....

   b Rainfall

   ....

   ....

3 CHANGING ECOSYSTEMS

The diagram shows a rainforest food web.

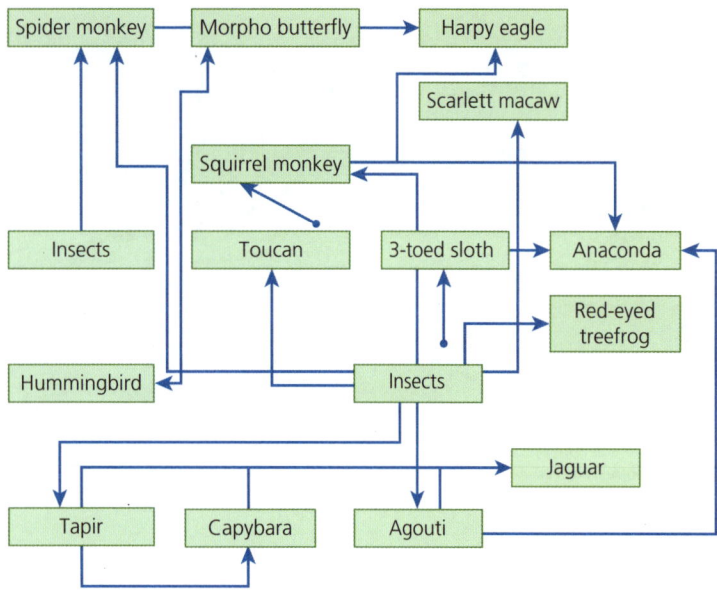

4 For areas of tropical rainforest, identify:

   a a primary producer

   ....

   b a herbivore

   ....

   c a carnivore

   ....

5 Suggest why Harpy eagles and jaguars depend on different food sources.

   ....

   ....

   ....

   ....

3.4 Threats to the tropical rainforest and how they can be managed

1 The diagram shows some of the threats to the Amazon rainforest. Briefly outline the impacts of:

a large-scale deforestation

....

....

....

b HEP projects

....

....

....

c new roads

....

....

....

d illegal logging

....

....

....

3 CHANGING ECOSYSTEMS

e poaching and the illegal wildlife trade

....

....

....

....

f overgrazing

....

....

....

....

g mining

....

....

....

....

h migration

....

....

....

....

i overfishing

....

....

....

2 Outline the main beneficiaries of managing the Amazon rainforest.

   ....
   ....
   ....
   ....
   ....
   ....

3 Evaluate ways in which it is possible to manage the Amazon rainforest.

   ....
   ....
   ....
   ....
   ....
   ....
   ....
   ....
   ....
   ....

4 Tectonic hazards

4.1 The structure of the Earth and distribution of earthquakes and volcanoes

1 Define the following terms:

   a crust

   ....

   b core

   ....

   c mantle

   ....

2 Describe what happens when continental crust meets oceanic crust.

   ....
   ....
   ....
   ....
   ....
   ....

3 Describe the distribution of the world's active volcanoes.

   ....
   ....
   ....
   ....
   ....
   ....

4 State **two** types of plate boundaries where volcanoes occur.

   ....
   ....

4.1 The structure of the Earth and distribution of earthquakes and volcanoes

5 Identify **one** type of plate boundary where volcanoes do not occur.

   ....

   ....

6 Explain what is happening at the Mid-Atlantic ridge that runs through Iceland.

   ....

   ....

   ....

   ....

   ....

   ....

4 TECTONIC HAZARDS

4.2 The processes and features associated with earthquakes and volcanoes

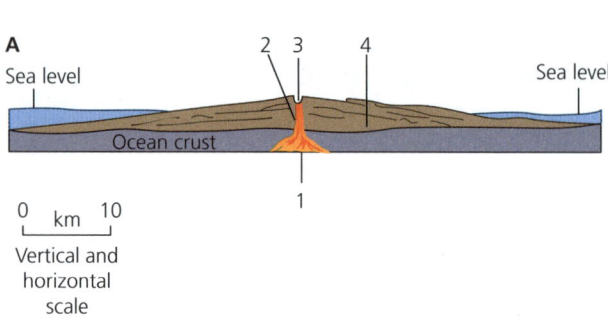
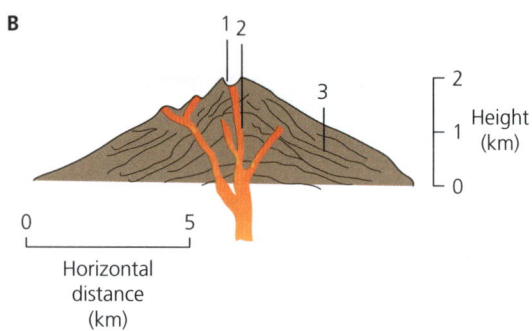

1 The figures show a cone volcano and a shield volcano.

   a Identify both types of volcano.

   A ....

   B ....

   b For each volcano, identify the numbered features.

   A

   1 ....

   2 ....

   3 ....

   4 ....

   B

   1 ....

   2 ....

   3 ....

   c Estimate the width and height of both types of volcano.

   A ....

   B ....

d Distinguish between magma and lava.

....

....

....

....

....

2 Describe the main characteristics of shield volcanoes and cinder cone volcanoes.

....

....

....

....

....

....

3 Define these volcano classifications:

   a active

   ....

   ....

   b dormant

   ....

   ....

   c extinct

   ....

   ....

4 TECTONIC HAZARDS

4 Describe **three** volcanic hazards.

4.3 The impacts of tectonic hazards

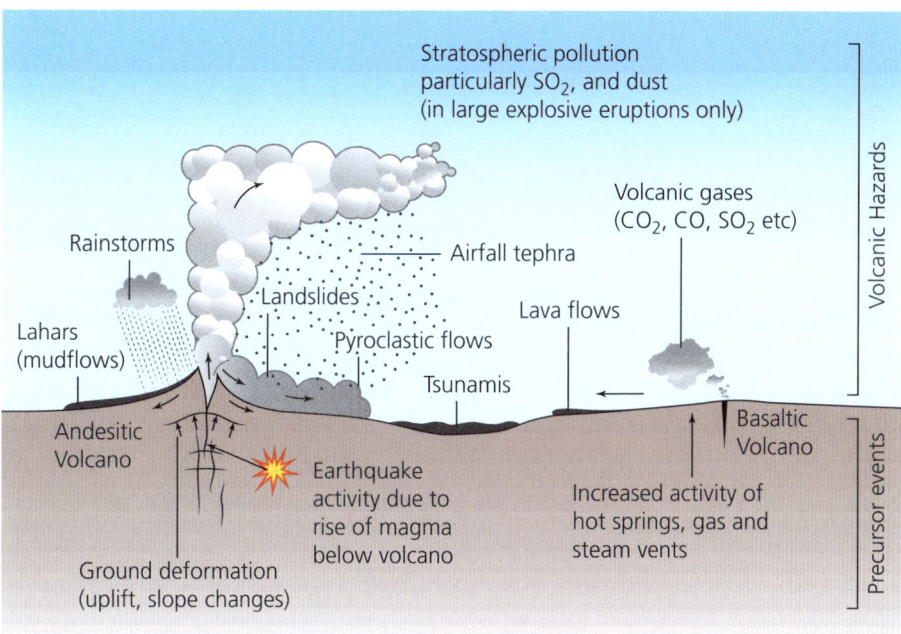

The diagram shows some of the changes that happen before a volcano erupts, and some of the impacts of a volcano.

1 a Define the term 'precursor events'.

   ....

   ....

   ....

  b Identify **three** precursor events that suggest a volcanic eruption is imminent.

   ....

   ....

   ....

2 a Identify the type of volcano formed by basaltic lava.

   ....

  b Identify the type of volcano formed by acidic lava.

   ....

  c State whether basaltic volcanoes or acidic volcanoes are more explosive.

   ....

4 TECTONIC HAZARDS

    d Explain the hazards caused by acidic and basaltic volcanoes.

    ....

    ....

    ....

    ....

    ....

3 Distinguish between lava flows and pyroclastic flows.

....

....

....

....

4 Define the terms 'ash fall' and 'lahar'.

Ash fall

....

Lahar

....

5 Suggest why volcanic eruptions vary in terms of speed.

....

....

....

....

6 Distinguish between the focus and the epicentre of an earthquake.

....

....

7 Outline **two** ways in which earthquakes can be measured.

....

....

....

4.3 The impacts of tectonic hazards

8 Identify **one** factor that influences the impact of earthquakes.

....

9 Explain **two** physical factors that influence the impact of earthquakes.

....

....

....

....

10 Explain **two** human factors that influence the impact of earthquakes.

....

....

....

....

4 TECTONIC HAZARDS

4.4 Managing the impacts of tectonic hazards

1 Examine the primary and secondary responses to tectonic hazards.

   ....

   ....

   ....

   ....

   ....

   ....

   ....

2 Evaluate the strategies and techniques used to manage the impacts of earthquakes.

   ....

   ....

   ....

   ....

   ....

   ....

   ....

3 Outline the causes and impacts of an earthquake on a named country/area you have studied.

   ....

   ....

   ....

   ....

   ....

   ....

   ....

4 Assess the responses to this earthquake.

   ....

   ....

   ....

   ....

4.4 Managing the impacts of tectonic hazards

....

....

....

5 Evaluate the strategies and techniques used to manage the impacts of volcanic eruptions.

....

....

....

....

....

....

....

6 Outline the causes and impacts of a volcano on a named country/area you have studied.

....

....

....

....

....

....

....

7 Assess the responses to this volcanic eruption.

....

....

....

....

....

....

....

4 TECTONIC HAZARDS

8 Evaluate the strategies and techniques used to manage the impacts of volcanic eruptions.

5 Climate change

5.1 The natural and human causes of climate change

The diagram shows atmospheric levels of CO_2 at the Mauna Loa Observatory, Hawaii, between 1958 and 2020.

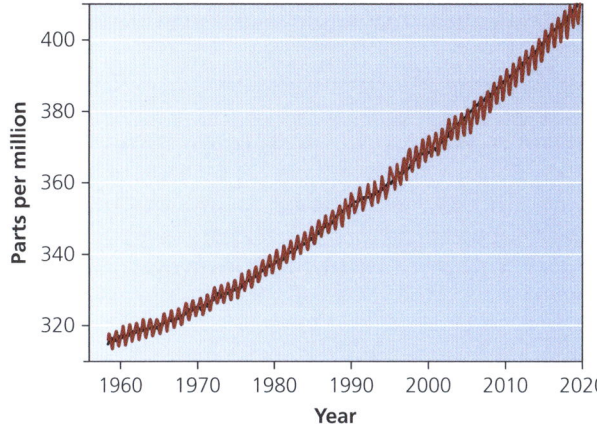

1   a   State the approximate concentration of CO_2 in 1958.

....

   b   State the approximate concentration of CO_2 in 2020.

....

   c   Calculate the absolute and relative increase of CO_2 between 1958 and 2020.

   Absolute ....

   Relative ....

   d   Suggest why the CO_2 readings are taken at Mauna Loa, Hawaii rather than elsewhere.

....

....

....

2   a CO_2 is a greenhouse gas. Explain what is meant by the term 'greenhouse gas'.

....

....

....

....

5 CLIMATE CHANGE

b Outline the difference between the greenhouse effect and the enhanced greenhouse effect (global warming).

....

....

....

....

....

....

....

....

5.2 The impacts of climate change at a range of geographical scales

The diagram shows historic CO_2 emissions by region, 1960–2018.

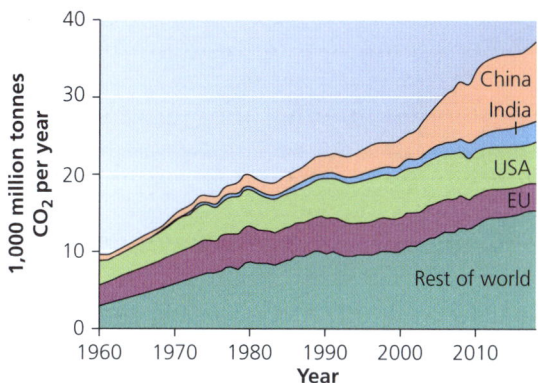

1 a Describe the change in total emissions between 1960 and 2018.

....

....

....

....

....

b Describe the changes in CO_2 emissions for (i) the EU, (ii) China and (iii) the USA.

  i EU

....

....

  ii China

....

....

  iii USA

....

....

2 a Briefly outline the physical impacts of increased CO_2 emissions.

....

....

....

....

5 CLIMATE CHANGE

b Suggest how the impacts of increased CO_2 emissions can be managed.

....

....

....

....

....

....

The diagram shows changes in (a) Northern Hemisphere spring snow cover, (b) Arctic summer sea ice extent and (c) global average sea level change between 1900 and 2020.

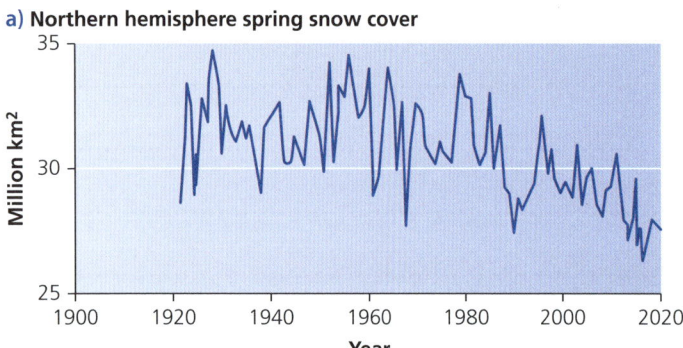

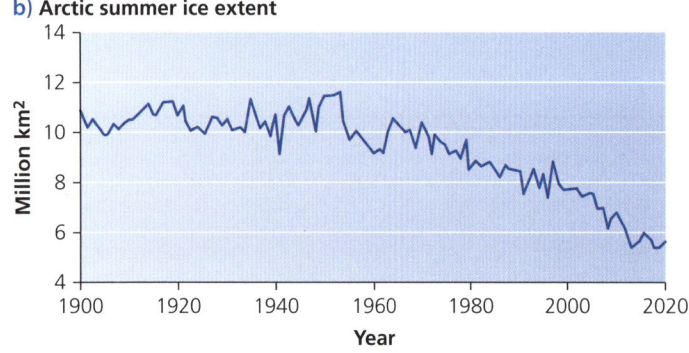

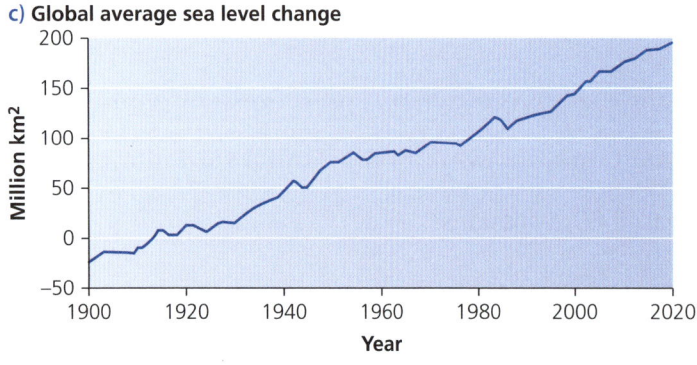

5.2 The impacts of climate change at a range of geographical scales

3 a Describe the change in Northern Hemisphere spring snow cover from 1900 to 2020.

....
....
....
....
....
....
....

b Analyse the changes in Arctic summer sea ice between 1900 and 2020.

....
....
....
....
....
....
....

c Briefly outline the changes in global average sea level between 1900 and 2020.

....
....
....
....
....

4 a Suggest **one** impact of the decline of snow cover and sea ice.

....
....
....
....
....
....

5 CLIMATE CHANGE

b Explain **two** reasons for sea levels to rise.

....

....

....

....

....

....

5.3 Responses to climate change

The diagram shows some of the ways that people can mitigate or adapt to climate change.

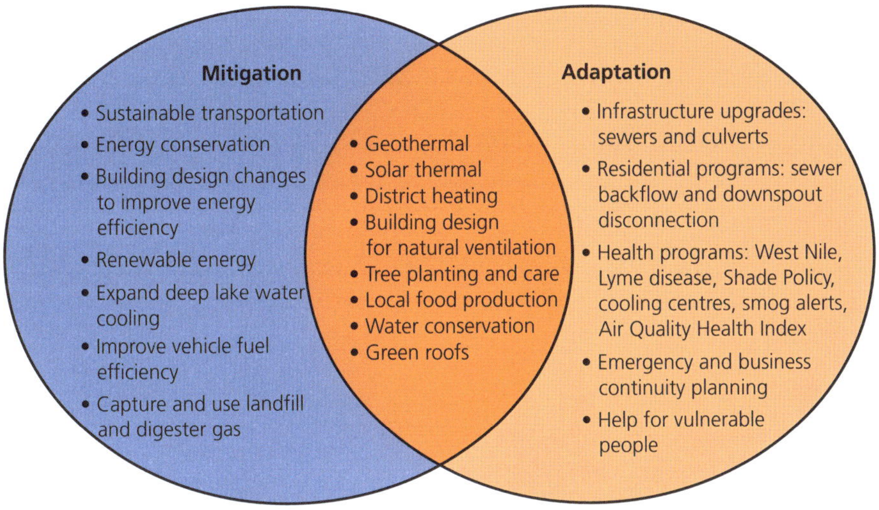

Mitigation
- Sustainable transportation
- Energy conservation
- Building design changes to improve energy efficiency
- Renewable energy
- Expand deep lake water cooling
- Improve vehicle fuel efficiency
- Capture and use landfill and digester gas

Both (overlap)
- Geothermal
- Solar thermal
- District heating
- Building design for natural ventilation
- Tree planting and care
- Local food production
- Water conservation
- Green roofs

Adaptation
- Infrastructure upgrades: sewers and culverts
- Residential programs: sewer backflow and downspout disconnection
- Health programs: West Nile, Lyme disease, Shade Policy, cooling centres, smog alerts, Air Quality Health Index
- Emergency and business continuity planning
- Help for vulnerable people

1 a Define 'climate mitigation'.

....

....

b Define 'climate adaptation'.

....

....

c State **two** methods of climate mitigation.

....

....

....

....

....

5 CLIMATE CHANGE

d State **two** methods of climate adaptation.

....

....

....

....

....

....

6 Changing populations

6.1 Populations grow and decline

Using the CIA World Factbook at **www.cia.gov/the-world-factbook/** go to the Countries tab at the top of the screen.

Click on any country and go to the Population section. Then scroll down and click on 'People and society'.

You can see on the CIA World Factbook website that looking at the population size (or other demographic aspects) for one country can also link to a list of the highest and lowest values for that indicator.

1 Using the CIA World Factbook, identify the **three** largest countries in the world by population.

   1 ....

   2 ....

   3 ....

2 Find out the population in the current year for:

   a Greenland ....

   b Lichtenstein ....

   c Montserrat ....

   d Holy See (Vatican City) ....

   e Antarctica

      i Summer ....

      ii Winter ....

3 Using an atlas, or other resources, suggest contrasting reasons for the small population sizes in the countries in Question 2.

   ....

   ....

   ....

   ....

   ....

   ....

   ....

6 CHANGING POPULATIONS

4 Define the terms 'birth rate', 'death rate' and 'fertility rate'.

Birth rate

....

....

....

Death rate

....

....

....

Fertility rate

....

....

....

5 State the units that birth rates, death rates and fertility rates are measured in.

Birth rate ....

Death rate ....

Fertility rate ....

6 a Identify the **three** countries with the highest birth rates.

Country	(‰)

b Identify the **three** countries with the lowest birth rates.

Country	(‰)

(NB You can omit St Pierre and Miquelon)

7 a Identify the **three** countries with the highest death rates.

Country	(‰)

b Identify the **three** countries with the lowest death rates.

Country	(‰)

8 Comment on the locations with the highest birth rates and death rates compared with the countries that have the lowest birth rates and death rates.

....

....

....

....

....

....

....

6 CHANGING POPULATIONS

6.2 Population structures change over time

The diagrams below show population pyramids for Japan in 1930, 1995 and 2025.

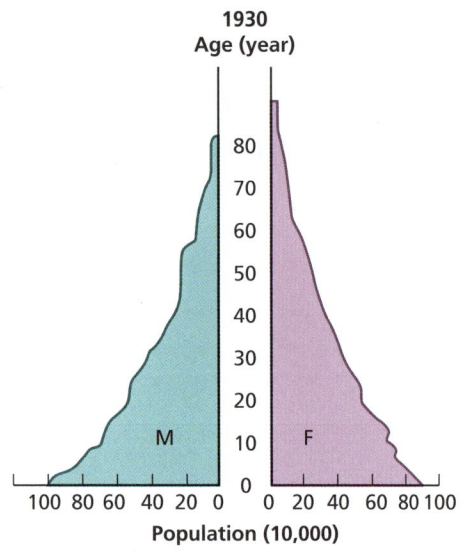

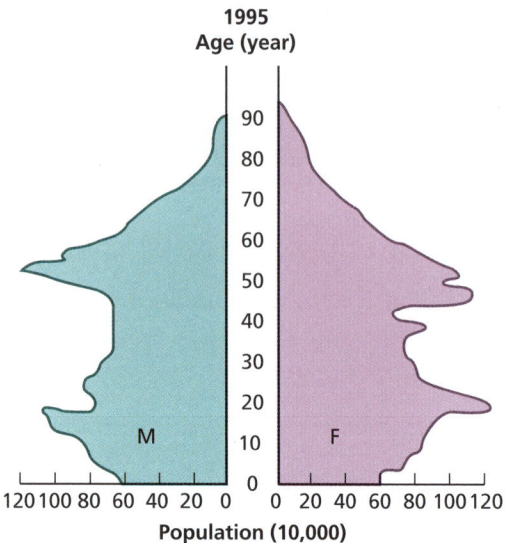

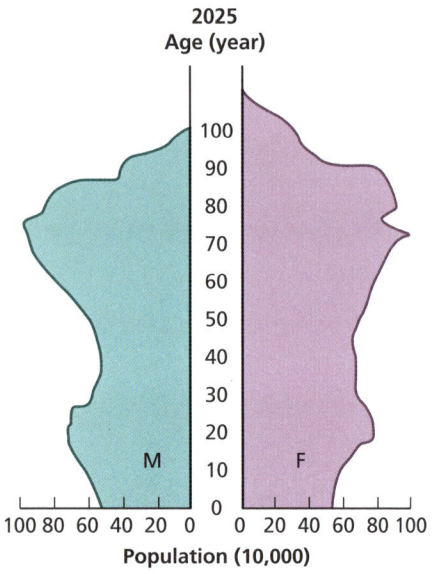

(Source: ResearchGate)

1 Describe the main changes in the population pyramids between 1930 and 2025.

....

....

....

....

....

....

6.2 Population structures change over time

2 Suggest **two** or more reasons for these changes.

   ....

   ....

   ....

   ....

3 Outline **two** or more problems caused by Japan's current population structure.

   ....

   ....

   ....

   ....

6 CHANGING POPULATIONS

6.3 The causes and impacts of international migration

The table shows the number of international migrants worldwide, 1970–2020.

Year	Number of international migrants (millions)
1970	84
1975	90
1980	101
1985	113
1990	152
1995	161
2000	173
2005	191
2010	220
2015	247
2020	280

(Source: Adapted from World Migration Report, 2024)

1 Plot the data for the number of international migrants 1970–2020.

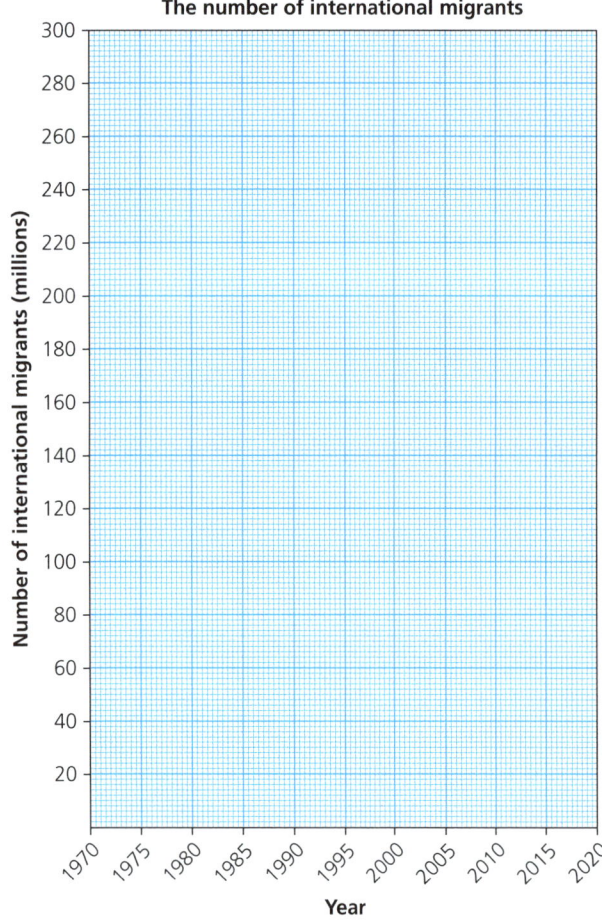

6.3 The causes and impacts of international migration

2 Calculate (a) the absolute increase and (b) the relative increase in the number of international migrants, 1970–2020.

   a Absolute increase = 2020 number of international migrants − 1970 number of international migrants

   ....

   ....

   b Relative increase = (2020 number of international migrants − 1970 number of international migrants)/1970 number of international migrants

   ....

   ....

3 Describe the changes in international migration between 1970 and 2020.

   ....

   ....

   ....

   ....

   ....

   ....

   ....

   ....

4 Suggest **two** reasons for the changes in international migration, 1970–2020.

   ....

   ....

   ....

   ....

   ....

   ....

   ....

6 CHANGING POPULATIONS

The diagram below shows international migration flows.

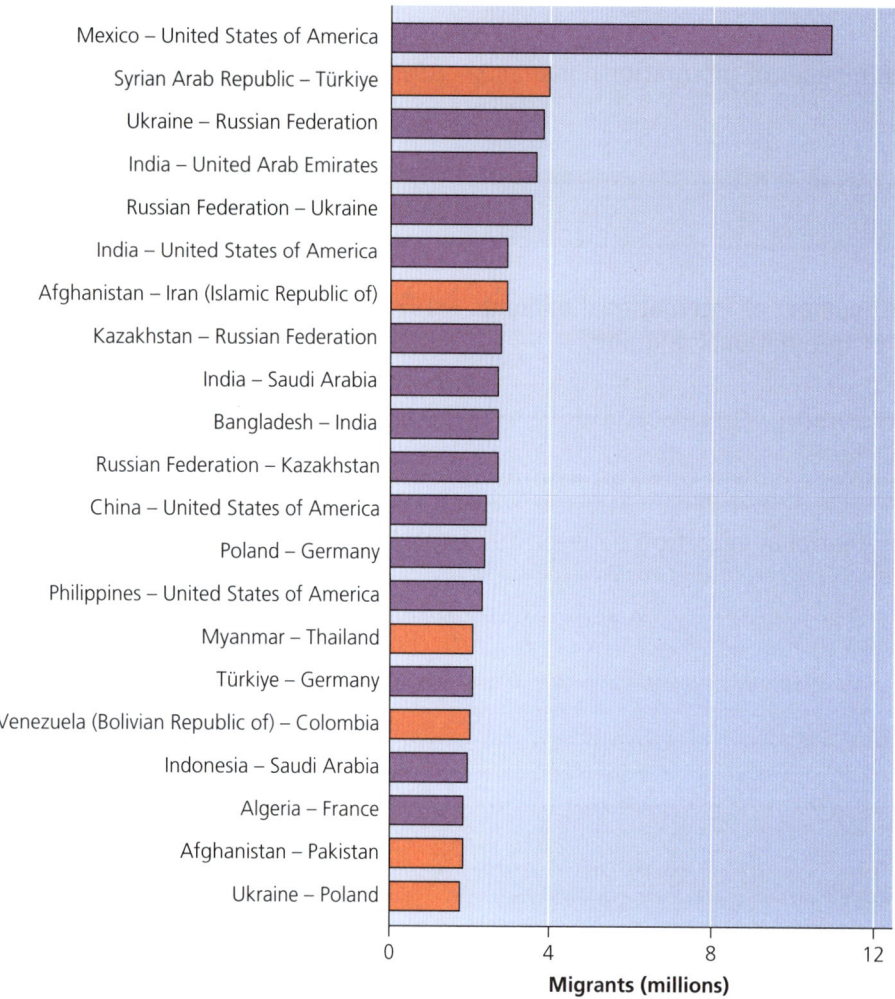

(Source: World Migration Report, 2024)

The bars/corridors show the origin (left) and destination (right) over the major migration flows.

The orange bars show mainly displaced people.

5 Identify the largest international migration flow and state its size.

....

6 Identify the main destinations of Ukrainian migrants.

....

....

7 Identify **two** examples in which migrants have moved to neighbouring countries.

....

....

6.3 The causes and impacts of international migration

8 Identify **two** examples in which migrants have moved to non-neighbouring countries.

   ....

   ....

9 Suggest what is meant by the term 'displaced persons'. State one example of a displaced population from South America and one from Europe.

   ....

   ....

   ....

10 Using an example country of your choice, explain how push and pull factors affect international migration.

   ....

   ....

   ....

   ....

11 Using your example country, outline the positive and negative impacts of international migration on the country of origin and the destination country. Write your answer on a separate piece of paper.

12 Using an example country, evaluate two strategies that have been used to manage international migration. Write your answer on a separate piece of paper.

7 Changing towns and cities

7.1 Where people live

The map below shows the estimated population in selected megacities, 2022–2030.

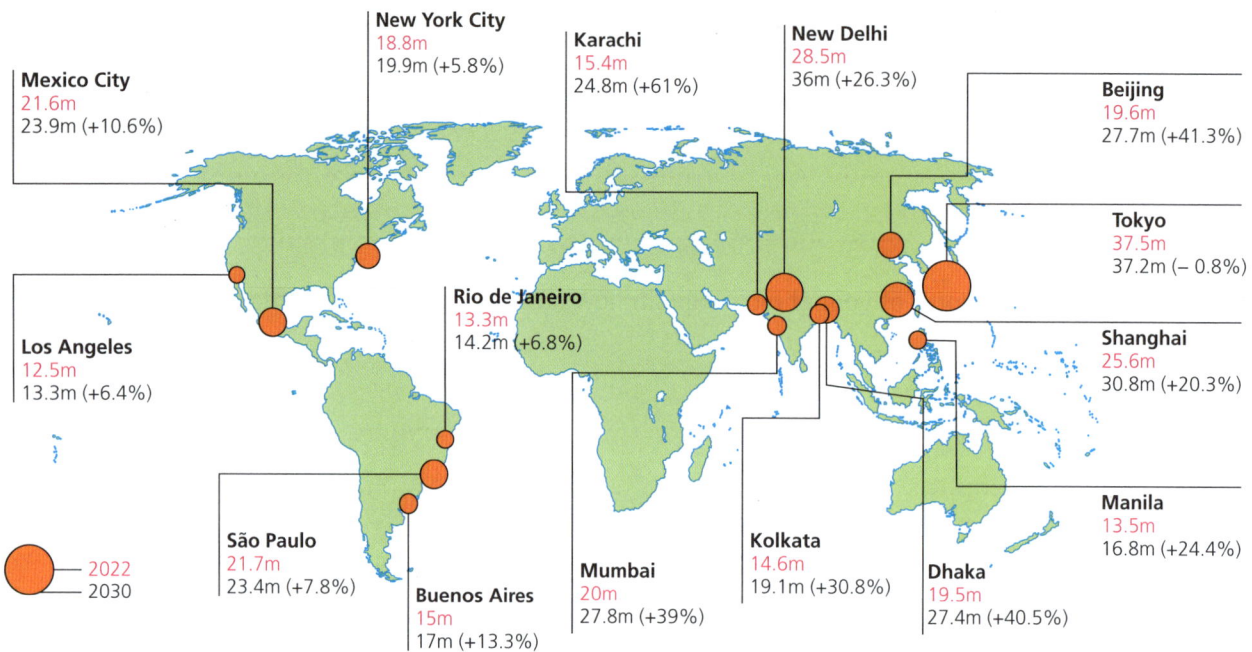

1 Distinguish between a millionaire city and a megacity.

 ....

 ....

2 a Identify the city that is predicted to grow most by 2030 in:

 i absolute size ....

 ii relative size..

 b Identify the city that is predicted to decline in size or the city that is expected to grow the least by 2030.

 ....

3 a Describe the global distribution of megacities as shown on the map.

 ....

 ....

 ....

7.1 Where people live

b Explain the causes of rapid urban growth in LICs.

....

....

....

....

....

....

c Explain how social and economic factors affect urban growth.

....

....

....

....

....

....

d Explain how environmental and political factors affect urban growth.

....

....

....

....

....

....

7 CHANGING TOWNS AND CITIES

7.2 The opportunities and challenges of urbanisation

The picture below shows suburban development in Oxford, UK.

1 Define the term 'urban sprawl'.

　...

　...

2 Outline factors that may lead to urban sprawl.

　...

　...

　...

　...

　...

　...

3 Describe **two** opportunities and **two** challenges of rapid urban growth.

　...

　...

　...

　...

　...

　...

4 Outline the impact of urban sprawl on the rural-to-urban fringe.

　...

　...

　...

　...

7.2 The opportunities and challenges of urbanisation

The photo below shows an out-of-town supermarket.

5 In the space around the photo, label the characteristics of the out-of-town supermarket shown.

6 Examine the opportunities of urban living.

   ....
   ....
   ....
   ....
   ....
   ....
   ....
   ....
   ....
   ....

7 Explain the issues of waste management in cities.

   ....
   ....
   ....
   ....
   ....
   ....

7 CHANGING TOWNS AND CITIES

The photograph shows part of the Cite Soleil district of Port au Prince, Haiti.

8 Define the term 'unplanned settlement'.

   ....

   ....

9 Describe the main characteristics of the Cite Soleil district of Port au Prince, as shown in the photograph.

   ....

   ....

   ....

   ....

   ....

   ....

   ....

10 Outline why migrants are attracted to places such as Cite Soleil.

   ....

   ....

   ....

   ....

   ....

   ....

   ....

7.2 The opportunities and challenges of urbanisation

11 Suggest some of the challenges that exist for residents of settlements such as Cite Soleil.

....
....
....
....
....
....
....

7 CHANGING TOWNS AND CITIES

7.3 The management of urban growth

The photo shows a 'green roof' in Singapore.

1 Suggest how the 'green roof' encourages urban sustainability.

   ....
   ....
   ....

2 Outline how transport in urban areas can be made more sustainable.

   ....
   ....
   ....
   ....
   ....
   ....
   ....

3 Identify ways in which housing in urban areas can be made more sustainable.

   ....
   ....
   ....
   ....
   ....

7.3 The management of urban growth

4 Outline the main obstacles to achieving urban sustainability.

....

....

....

....

....

....

8 Development

8.1 Measuring development

1 a Define the terms GDP, GNP and GNI.

   GDP

   ....

   ....

   GNP

   ....

   ....

   GNI

   ....

   ....

The map shows global variations in GNI per person.

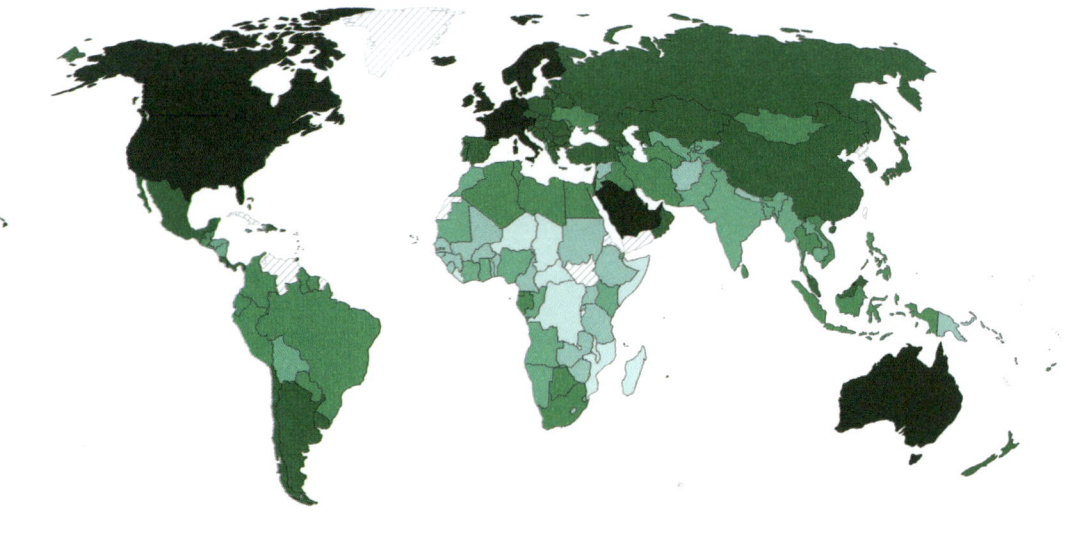

(Source: Our World in Data)

8.1 Measuring development

2 a State the level of GNP in:

  i Australia ....

  ii UK ....

  iii China ....

  iv India ....

  v Ethiopia ....

b Describe the global variations in GNI per person.

....

....

....

....

....

The table shows variations in literacy rates around the world. The global average in 2024 was 86.81.

Highest		Lowest	
Country	Literacy rate	Country	Literacy rate
Andorra	100.00	Chad	26.76
Finland	100.00	Mali	30.76
Greenland	100.00	South Sudan	34.52
Liechtenstein	100.00	Botswana	36.75
Luxembourg	100.00	Afghanistan	37.27
Norway	100.00	Niger	37.34
Ukraine	100.00	Central African Republic	37.49
Uzbekistan	100.00	Somalia	37.80
North Korea	100.00	Guinea	45.33
Latvia	100.00	Benin	45.84

(Source: US Career Institute, 2024)

3 a State the main characteristics of countries with very high levels of literacy.

....

....

....

b State the main characteristics of countries with very low levels of literacy.

....

....

....

8 DEVELOPMENT

4 Briefly outline the importance of literacy for development.

....

....

....

....

This table shows the HDI in a variety of different countries in 2010 and 2022.

Human development descriptor (HDI)	Rank (out of 192)	Country	2010	2022
Very high	1	Switzerland	0.940	0.967
High	70	Bulgaria	0.790	0.799
Medium	119	Venezuela	0.759	0.699
Low	161	Nigeria	0.485	0.548
Low	192	South Sudan	0.406	0.381

5 a Calculate the change in HDI between 2010 and 2022 for each of the countries in the table.

  i Switzerland ....

  ii Bulgaria ....

  iii Venezuela ....

  iv Nigeria ....

  v South Sudan ....

b State the country whose HDI (i) increased the most and (ii) decreased the most.

  i ....

  ii ....

This diagram shows dimensions, indicators and indexes related to the human development index (HDI).

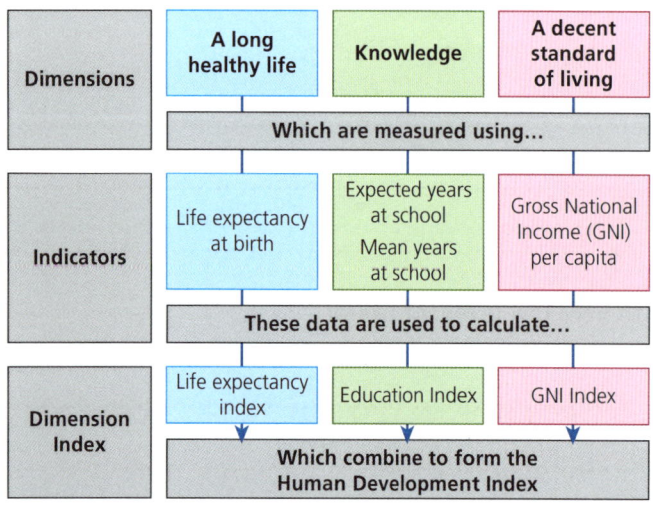

[Source: HDI is based on https://academistan.com/human-development-index-hdi-concept-and-history/]

6 a Identify the **three** dimensions of the HDI.

1 ....

2 ....

3 ....

b State the **four** indicators combined to produce the HDI.

1 ....

2 ....

3 ....

4 ....

7 The HDI is a composite indicator of development. Explain the meaning of the term 'composite' and explain why the HDI is a good measurement of level of development.

....

....

....

....

....

....

The table below shows the level of income and the infant mortality rate (IMR) for ten countries.

Country	Level of income (US $)	IMR (‰)	Rank (GNP)	Rank (IMR)	Difference (d) in ranks	Difference² (d²)
Cambodia	5 100	27.9				
Costa Rica	25 800	6.7				
Egypt	17 000	16.8				
Germany	61 900	3.1				
India	9 200	30.4				
Japan	46 300	1.9				
Niger	1 600	64.3				
Singapore	127 500	1.5				
Türkiye	34 400	18.4				
Vietnam	13 700	14.1				

(Source: CIA World Factbook)

8 DEVELOPMENT

The scattergraph below shows the plot of the first two countries.

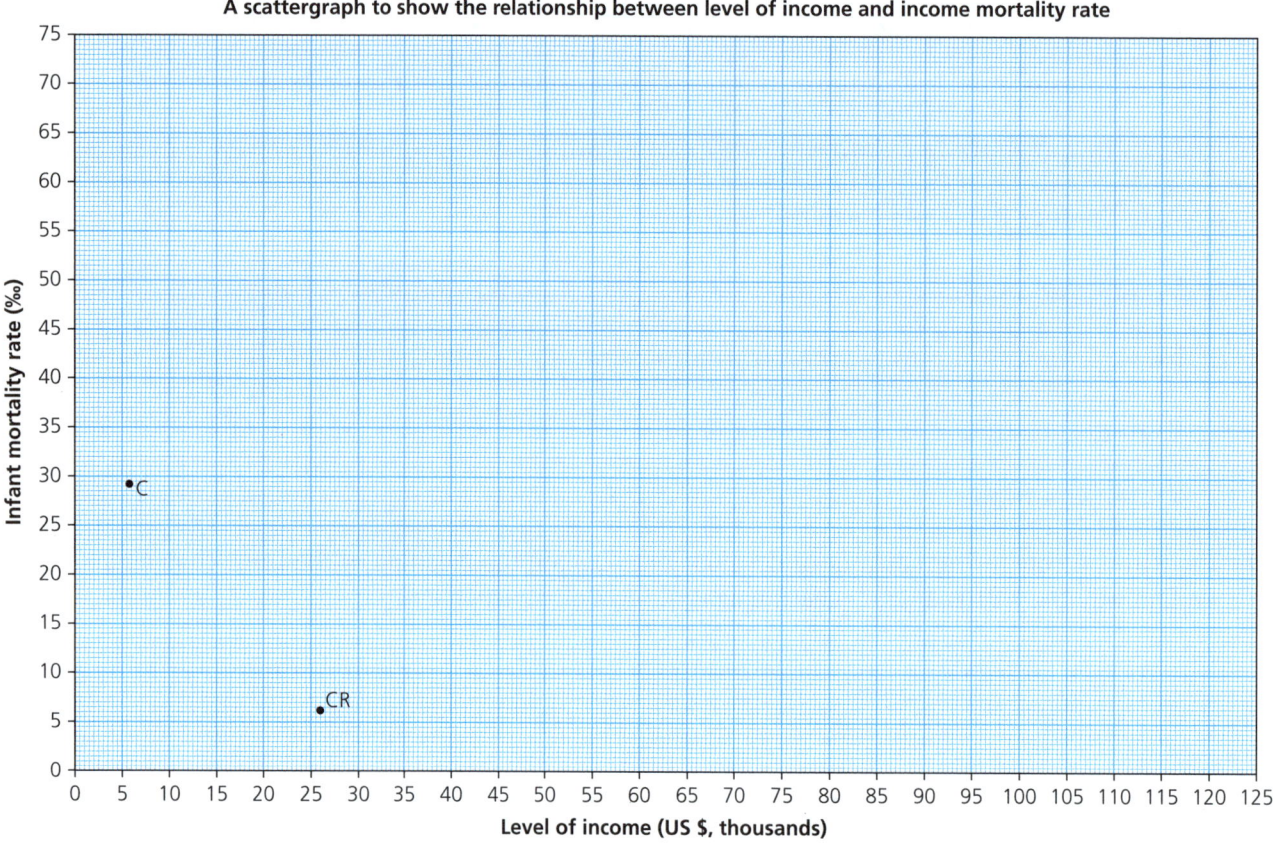

8 a Complete the scattergraph.

b Add a line of best fit.

c Describe the relationship between level of income and infant mortality rate, as shown on the graph that you have drawn.

....

....

To test whether there is a statistically significant statistical relationship between the two sets of data, we can use a Spearman's rank correlation coefficient test. This is found by using the formula:

$Rs = 1 - (6\sum d^2/n^3 - n)$

where d is the difference in ranks and n the number of observations.

It looks difficult at first, but if you follow the steps, it is relatively easy to compute and assess.

- First, complete the table, ranking the data for GNP and IMR in order of highest (rank 1) to lowest (rank 10).
- Find the difference (d) in ranks for each country. The sign (+/-) does not matter as squaring the difference will remove any negative numbers.
- Square the difference in ranks to find (d^2).
- Add up all the values of (d^2) to produce $\sum d^2$.

- Use the formula $R_s = 1 - (6\sum d^2/n^3 - n)$.
- Spearman's rank will produce a value of between -1 and +1. A positive result suggests that there is a positive relationship between two variables – as one factor increases, so too does the other. In contrast, a negative result suggests that as one value increases the other decreases.
- For a data set of ten values (as in this case) a value of over 0.56 (check) is over 95% statistically significant, whereas a value of over 0.75 is 99% statistically significant.

The map shows variations in global calorie intake.

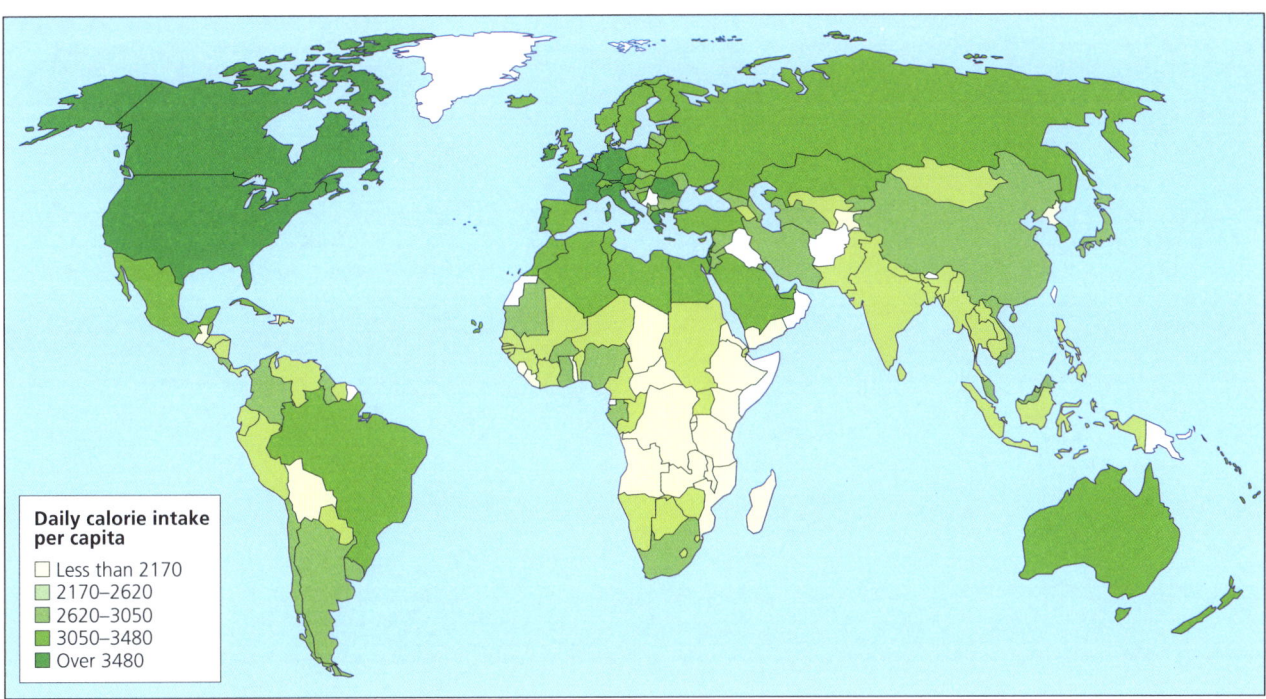

9 Identify **one** country with a calorie intake of (i) over 3480 calories per day and (ii) one with less than 2170 calories per day.

   i ....

   ii ....

10 Describe the main pattern of global calorie intake as shown on the map.

   ....

   ....

   ....

   ....

11 How does low calorie intake impact quality of life/standard of living?

   ....

   ....

   ....

   ....

8 DEVELOPMENT

12 Suggest **two** ways in which the problems caused by low calorie intake could be addressed.

....

....

....

....

8.2 The world is developing unevenly

1 Define the term 'development gap'.

   ....

   ....

2 State **one** social, **one** economic and **one** environmental factor that influences a country's level of development.

   Social ....

   Economic ....

   Environmental ....

The map shows the distribution of different types of countries according to the World Bank classification. The bar charts below show the number of each type of country by continent.

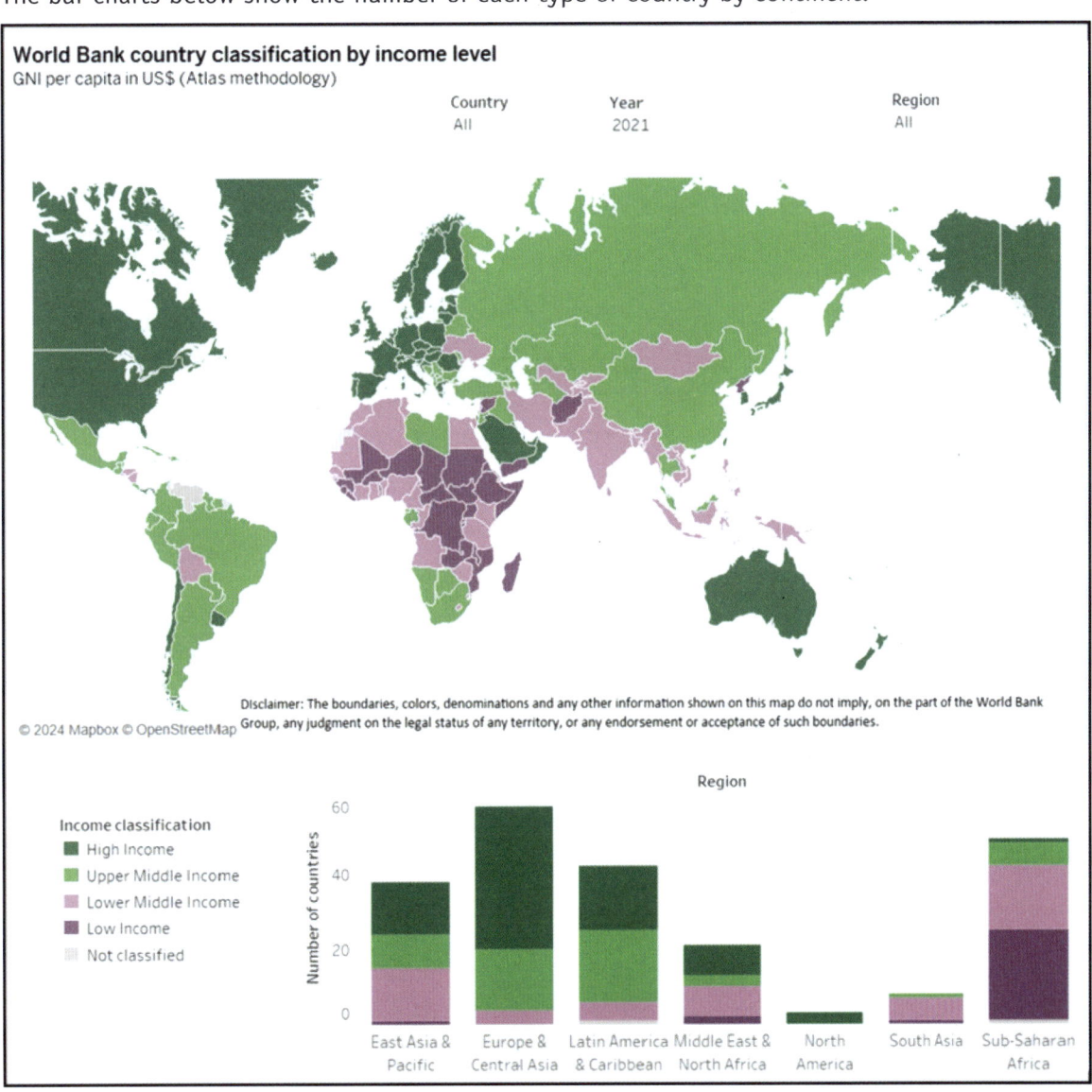

(Source: Our World in Data, 2024)

8 DEVELOPMENT

3 Using the internet (**https://blogs.worldbank.org/en/opendata/world-bank-country-classifications-by-income-level-for-2024-2025**) or other source, state the income levels of:

a high-income countries ....

b middle-income countries ....

c low-income countries ....

4 a Describe the distribution of high-income countries.

....

....

....

b Describe the distribution of low-income countries.

....

....

....

5 a Which continent has (i) the highest absolute number of HICs and (ii) the highest relative number of HICs.

i Absolute ....

ii Relative ....

b Which continent has (i) the highest absolute number of LICs and (ii) the highest relative number of LICs.

i Absolute ....

ii Relative ....

6 Using an atlas or other resource, identify:

a the LIC in East Asia and the Pacific

....

b one LIC in the Middle East and North Africa

....

8.3 Achieving sustainable development

In 2015, the UN announced the Sustainable Development Goals (SDGs), a collection of 17 goals that aimed to improve the quality of life among humans and to protect the planet.

Society	Economy	Environment
1 Poverty	7 Energy systems	13 Climate change
2 Hunger	8 Work and economic growth	14 Water ecosystems
3 Health and well-being	9 Industry, innovation and infrastructure	15 Land ecosystems
4 Education	10 Reduce inequalities	16 Peace, justice and strong institutions
5 Gender equality	11 Sustainable cities and communities	
6 Clean water and sanitation	12 Responsible consumption and production	17 Partnerships for the goals

1 Define the term 'sustainable development'.

   ....

   ....

2 a State the sustainable development goal that is concerned about the provision of clean water.

   ....

   b Identify **three** or more SDGs that are indirectly connected to the provision of clean water.

   ....

   ....

   ....

3 a Suggest **one** advantage of the SDGs.

   ....

   ....

   ....

   b Explain **one** disadvantage of the SDGs.

   ....

   ....

   ....

4 For a country of your choice, identify which of the SDGs you think is (i) the most important and (ii) of least importance. Justify your answers.

   i ....

   ....

   ....

   ....

8 DEVELOPMENT

ii ....

....

....

....

The diagram below shows trade between a high-income country and a low-income country. Describe the main features of the pattern of trade between the two types of country.

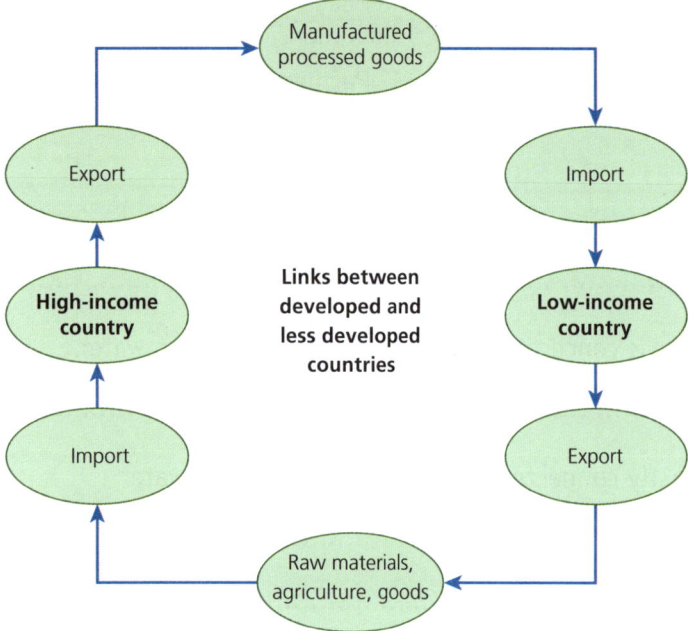

5 Suggest **two** advantages and **two** disadvantages of using trade as a strategy to reduce uneven development.

....

....

....

....

The table below shows the world's largest exporters, 2019.

Country	% of total world exports (goods, services and income)
USA	12.7
China	10.1
Germany	7.2
Japan	4.2
UK	4.0
France	3.8
Netherlands	3.8
Italy	2.5
South Korea	2.4
Canada	2.3

(Source: Data adapted from The Economist Pocket World in Figures, 2022)

8.3 Achieving sustainable development

6 a Plot the data in the grid below.

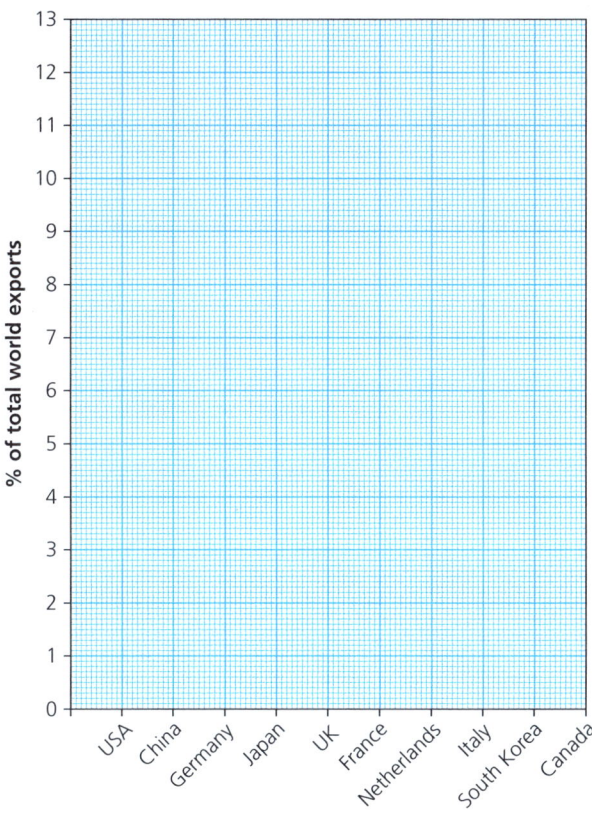

b Describe the main features of the world's top ten exporters. Explain why the world's top ten exporters are mostly HICs.

....

....

....

....

The diagram below shows bilateral aid, multilateral aid and charities.

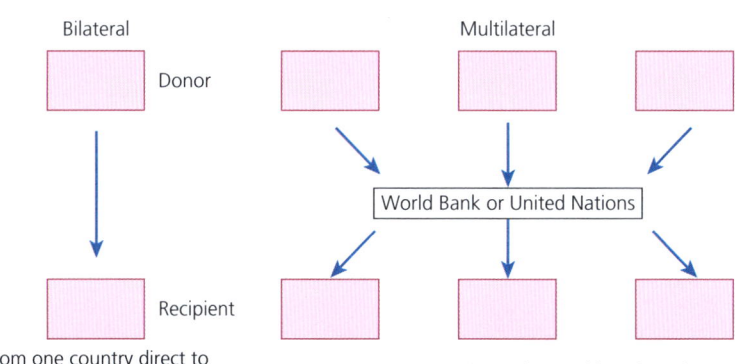

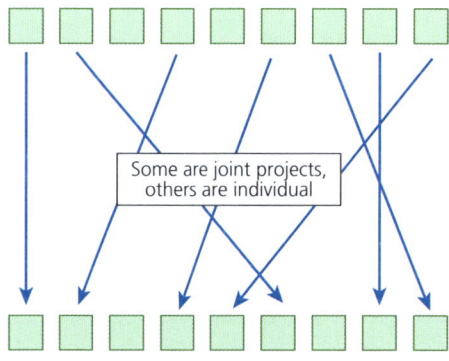

8 DEVELOPMENT

7 Define bilateral and multilateral aid.

Bilateral

....

....

Multilateral

....

....

The table below shows the world's largest donors, in total amount donated and in percentage of GNP.

Country	Amount donated (US $ bn)	Donation as a % of GNP
USA	33.5	0.2
Germany	24.2	0.6
UK	19.4	0.7
Japan	15.6	0.3
France	12.2	0.4
Turkiye	8.7	1.2
Netherlands	5.3	0.6
Sweden	5.2	1.0
Canada	4.7	0.3
Italy	4.4	0.2

8 a Plot the data for the world's largest donor countries on the graph below.

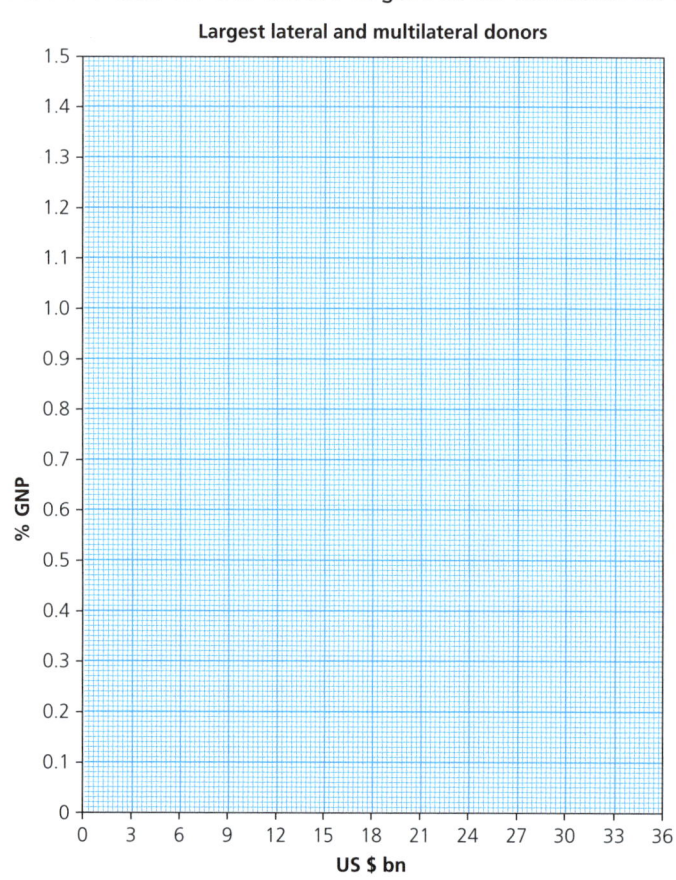

Largest lateral and multilateral donors

b Calculate Spearman's rank correlation coefficient (Spearman's rank = $1 - 6 \sum d^2/(n^3 - n)$) to test whether there is a statistically significant relationship between total donation and percentage of GNP.

Country	Amount donated (US$ bn)	Donation (% of GNP)	Rank (US$ bn)	Rank (% of GNP)	Difference in ranks	Difference²
						$\sum$

9 a Evaluate the impact of aid as a strategy to reduce uneven development.

....

....

....

....

....

....

....

b Compare the main donors of international aid and the main recipients.

....

....

....

....

....

....

....

9 Changing economies

9.1 Changing employment structures

1 There are four main types of industries: primary, secondary, tertiary and quaternary.

Four definitions are given below. Match the definition to the industrial type.

A	Service industries such as retailing, healthcare, education and finance.
B	High technology, including research and development.
C	The manufacture of raw materials into finished and semi-finished goods.
D	The harvesting of resources such as forestry, fishing, mining and farming.

....

....

....

....

2 The data below shows the employment structure for selected countries.

Country	Primary industries	Secondary industries	Tertiary industries
Bangladesh	38	21	41
China	25	28	47
France	3	20	77
India	43	25	32
Ivory Coast	40	13	47
Japan	3	24	73
Nigeria	35	12	53
Saudi Arabia	2	25	73
UAE	2	34	64
UK	1	18	81
USA	1	20	79

State the country that has the highest proportion of the workforce employed in:

a primary industries ....

b secondary industries ....

c tertiary industries ....

3 Plot the employment structure of the countries listed in the table above on the triangular graph.

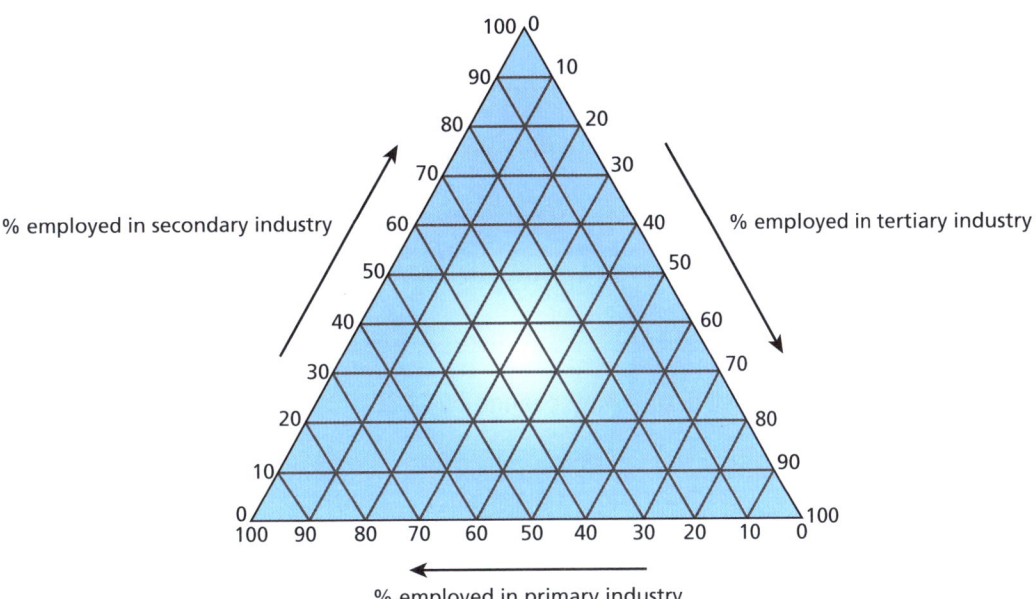

The data below shows changes in China's economic structure, 1990–2023.

	Primary	Secondary	Tertiary
1990	60	21	19
1995	52	23	25
2000	50	23	27
2005	45	24	31
2010	37	29	35
2012	34	30	36
2019	25	28	47
2023	23	29	48

4 a Plot three separate lines for each type of China's employment structure on the grid below.

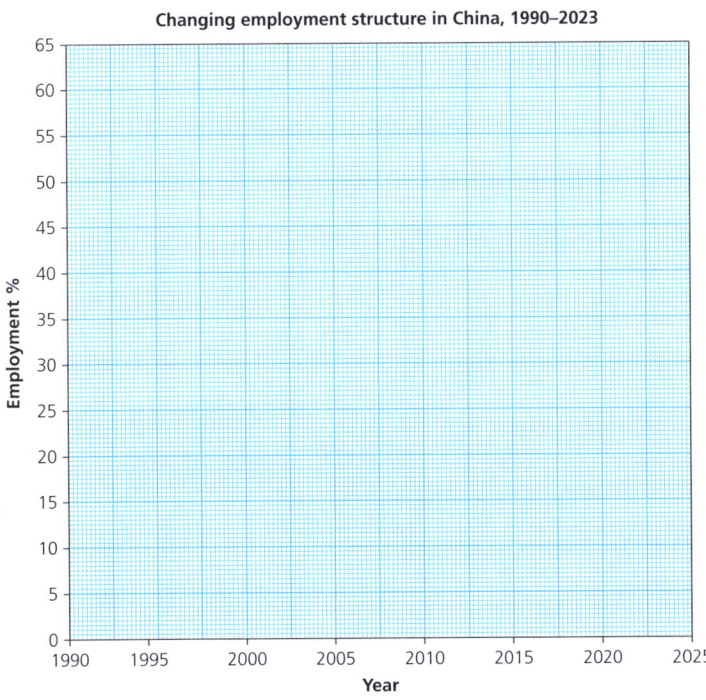

Changing employment structure in China, 1990–2023

9 CHANGING ECONOMIES

   b Describe the changes in China's economic structure between 1990 and 2023.

   ....

   ....

   ....

   ....

   ....

   ....

   c Predict how the employment structure may change by 2030.

   ....

   ....

   ....

   ....

   ....

   ....

   d Suggest **one** reason why employment in quaternary industries is not shown on the diagram.

   ....

   ....

5 Explain the influence of the following factors on the location of industry:

   a land

   ....

   ....

   b labour

   ....

   ....

   c raw materials

   ....

   ....

d energy

....

....

e transportation

....

....

f markets

....

....

g political policies

....

....

h technology

....

....

i communications

....

....

j containerisation

....

....

9 CHANGING ECONOMIES

9.2 The impact of globalisation and the role of transnational corporations

1 Define the term 'globalisation'.

   ....

   ....

   ....

2 Describe the key features of globalisation.

   ....

   ....

   ....

   ....

   ....

   ....

3 Outline **one** impact of globalisation on each of the following:

   a trade

   ....

   ....

   b transport

   ....

   ....

   c culture

   ....

   ....

   d communications

   ....

   ....

9.2 The impact of globalisation and the role of transnational corporations

e technology

....

....

4 Briefly explain the global organisation of transnational corporations (TNCs).

....

....

....

....

....

5 Evaluate the positive and negative impacts of TNCs on the countries in which they are located.

....

....

....

....

....

....

....

....

....

....

....

....

....

....

9 CHANGING ECONOMIES

9.3 Tourism is a growing industry

1 Identify **one or more** economic, social and political factor(s) that have led to the growth of the tourism industry.

Economic factors

....

....

Social factors

....

....

Political factors

....

....

2 Describe the evolution of a tourist resort with reference to the Butler model, as shown in the diagram below.

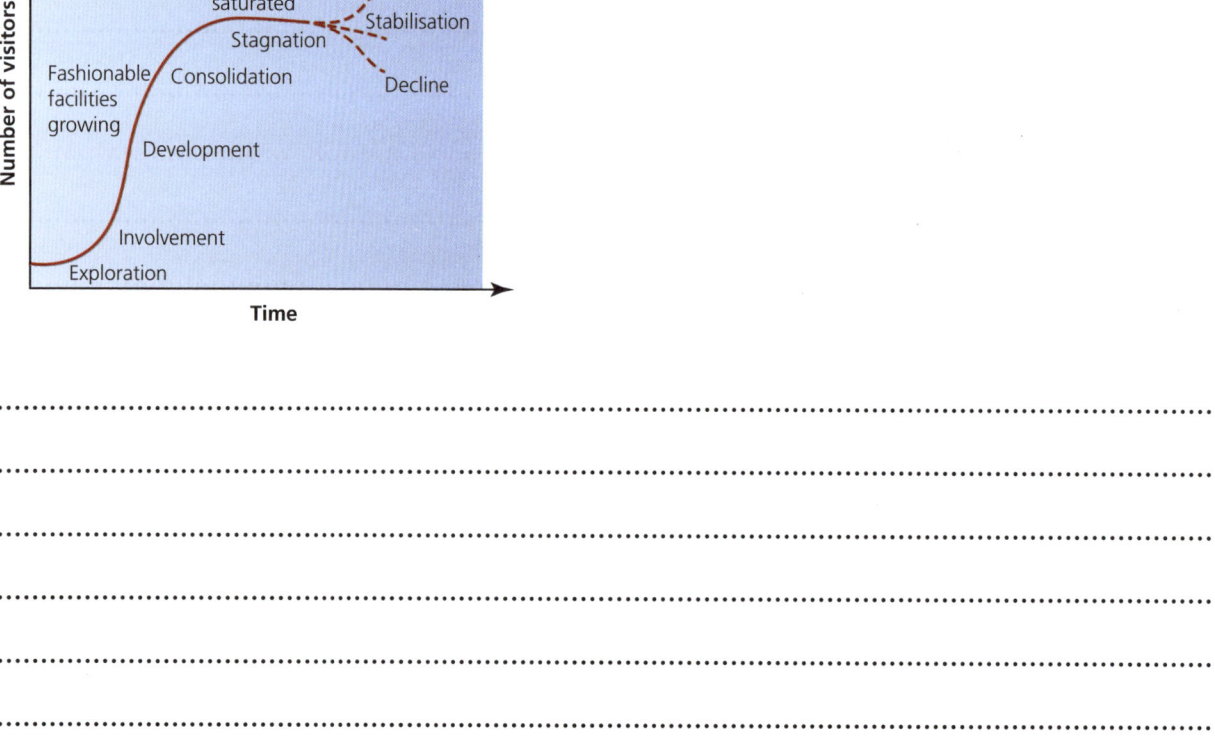

....

....

....

....

....

....

9.3 Tourism is a growing industry

3 Identify **three** benefits and **three** problems caused by tourism at a variety of scales.

....

....

....

....

....

....

....

4 a For a country that you have studied, identify **two or more** strategies to manage tourism.

....

....

....

....

....

....

....

b For a country that you have studied, assess the effectiveness of strategies to manage the impacts of tourism.

....

....

....

....

....

....

....

10 Resource provision

10.1 How food is produced

1 Define the following farming types: subsistence, commercial, arable, pastoral, aeroponics, aquaponics/hydroponics.

Subsistence

....

Commercial

....

Arable

....

Pastoral

....

Aeroponics

....

Aquaponics

....

Hydroponics

....

2 Explain, using an example, how farming systems are comprised of inputs, processes and outputs. A diagram could be used to explain an agricultural system.

....

....

....

10.2 Global patterns of food supply and demand

1 The tables show global corn production (2019) and rice production (2019).

Corn

Country	2019 Production (million tonnes)	% of global production
USA	347.0	30.2
China	260.8	22.7
Brazil	101.1	8.8
Argentina	56.9	4.9
Ukraine	35.9	3.1

Rice

Country	2019 Production (million tonnes)	% of global production
China	209.6	27.7
India	177.7	23.5
Indonesia	54.6	7.2
Bangladesh	54.6	7.2
Vietnam	43.5	5.8

1 a Compare the main producers of corn with those of rice.

....

....

....

....

b Suggest **two or more** reasons why corn and rice are grown in different countries/regions.

....

....

....

....

c Suggest reasons for changes in global food consumption.

....

....

....

....

....

....

....

10 RESOURCE PROVISION

d i Identify **two** strategies and techniques to increase food supply.

....

....

....

....

....

ii Discuss the effectiveness of these strategies to increase food supply.

....

....

....

....

....

e State **two** reasons for the globalisation of food supplies.

....

....

....

....

....

....

....

f Explain **two** impacts of the globalisation of food supplies.

....

....

....

....

....

....

....

10.3 The challenges of food supply

1 Assess **two** human and **two** natural factors that negatively affect food supply.

   Human factors

   ....

   ....

   ....

   ....

   Natural factors

   ....

   ....

   ....

   ....

2 Explain **three** causes of food insecurity.

   ....

   ....

   ....

   ....

   ....

   ....

   ....

   ....

   ....

3 Describe **two** problems caused by food insecurity for a named LIC, MIC or HIC.

   ....

   ....

   ....

   ....

   ....

   ....

10 RESOURCE PROVISION

4 Describe **two** advantages and **two** disadvantages of food aid.

....

....

....

....

....

....

....

....

....

....

The map shows areas at risk of desertification.

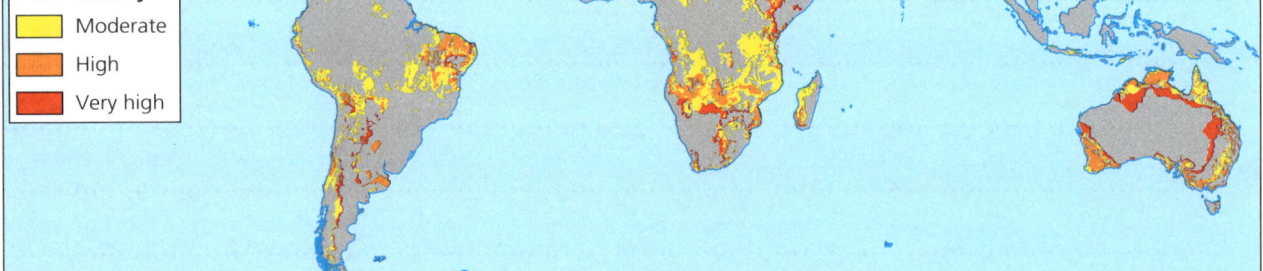

(Source: data from US Department of Agriculture)

5 a Define the term 'desertification'.

....

....

....

....

b Describe the distribution of areas at (i) very high risk and (ii) high risk of desertification.

  i ....

  ....

  ii ....

  ....

c Outline **two** causes of desertification.

....

....

....

....

....

d Describe **two** impacts of desertification.

....

....

....

....

....

....

....

e Explain **two** ways in which desertification can be managed.

....

....

....

....

....

....

....

10 RESOURCE PROVISION

10.4 How our energy is produced

1 Define 'renewable energy'.

....

....

2 Define these types of renewable energy:

  a biomass

  ....

  ....

  b geothermal

  ....

  ....

  c hydro-electric power (HEP)

  ....

  ....

  d solar

  ....

  ....

  e tidal

  ....

  ....

  f wave

  ....

  ....

  g wind

  ....

  ....

3 Define 'non-renewable energy'.

  ....

  ....

10.4 How our energy is produced

4 Define these types of non-renewable energy:

   a fossil fuels

....

....

....

   b nuclear power

....

....

....

5 Define 'fuelwood'.

....

....

....

10 RESOURCE PROVISION

10.5 The global patterns of energy supply and demand

The graph shows global primary consumption by source.

Global primary energy consumption by source

Primary energy is calculated based on the 'substitution method' which takes account of the inefficiencies in fossil fuel production by converting non-fossil energy into the energy inputs required if they had the same conversion losses as fossil fuels.

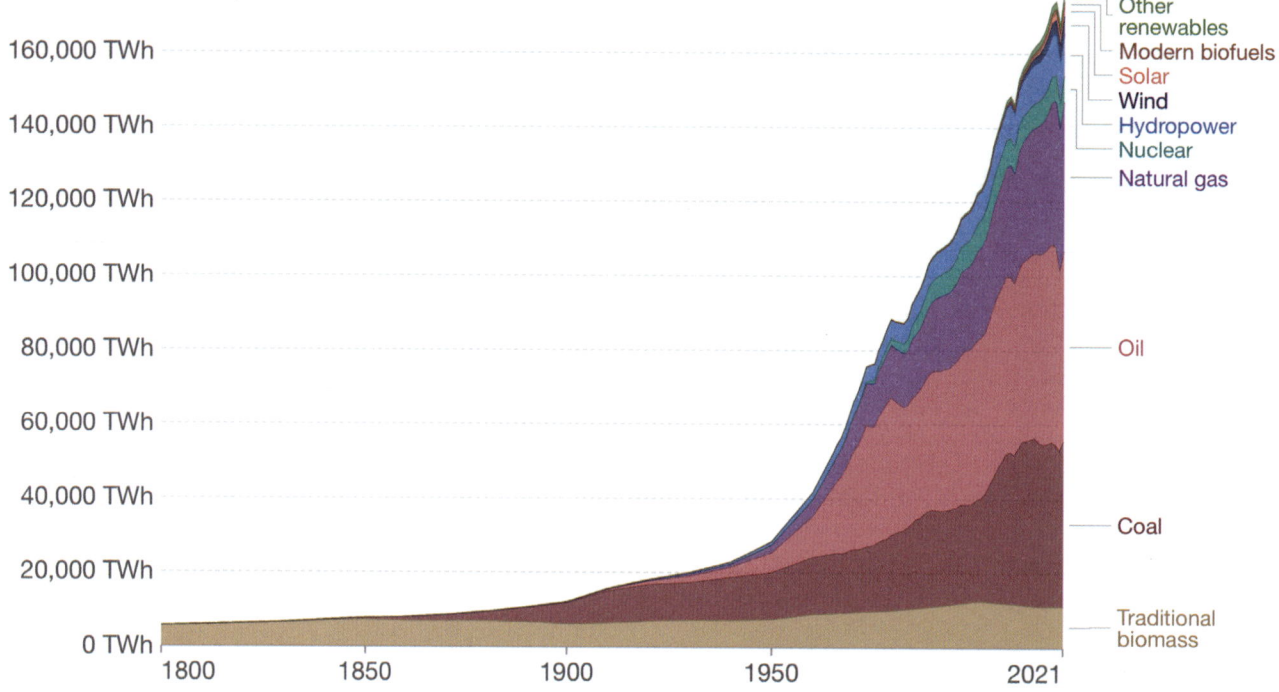

Source: Our World in Data based on Vaclav Smil (2017) and BP Statistical Review of World Energy   OurWorldInData.org/energy • CC BY

1 a Describe the changes in the total energy consumption between 1800 and 2021.

....

....

....

b Describe the changes in energy consumption by type of energy between 1800 and 2021.

....

....

....

....

....

....

10.5 The global patterns of energy supply and demand

c Briefly explain the reasons for the increasing global consumption of energy.

....
....
....
....
....
....

The diagram shows regional consumption of fuel, 2018.

Regional consumption by fuel, 2018
Percentage

Key:
- Coal
- Renewables
- Hydroelectricity
- Nuclear energy
- Natural gas
- Oil

Regions: North America, S. & Cent. America, Europe, CIS, Middle East, Africa, Asia Pacific

2 a Compare the regional consumption of fuel in the Middle East and Europe.

....
....
....
....

b Describe the global patterns of energy surplus and deficit and the importance of energy security.

....
....
....
....
....
....

10 RESOURCE PROVISION

10.6 The impacts of energy production

1 Describe **two** advantages and **two** disadvantages of one renewable energy source and one non-renewable source.

Renewable source:

....

....

....

....

Non-renewable source:

....

....

....

....

2 Describe **two** techniques that could increase the energy supplies of a country or area.

....

....

....

....

....

....

....

....

Energy sources in Nepal

Nepal's energy sources are biomass (75 per cent) and petroleum products (22 per cent) Renewable sources account for about three per cent. Nepal has no major oil or coal resources. The country is remote, mountainous and land-locked making imports difficult and expensive. Biomass including fuelwood, agricultural waste and dung is the main form of energy in Nepal. Nepal has huge potential for hydroelectric power. There are over 120 HEP plants in use presently, and plans for a further 250. However, the Nepal earthquake of 2015 destroyed at least 14 HEP plants affecting c. 30 per cent of Nepal's generating capacity. This prompted the government to try and diversify Nepal's energy mix. It plans to develop solar power.

10.6 The impacts of energy production

3 a Describe Nepal's energy mix.

   ....
   ....
   ....
   ....

   b Identify **two** forms of renewable energy in Nepal.

   ....
   ....

   c Outline **two** problems associated with Nepal's use of renewable energy.

   ....
   ....
   ....
   ....

   d Explain **two** reasons why it may be difficult for Nepal to rely on fossil fuels.

   ....
   ....
   ....
   ....

4 Evaluate strategies and techniques to manage energy supplies.

   ....
   ....
   ....
   ....
   ....
   ....
   ....
   ....

Geographical skills

1 State the names by which the following lines of latitude are known and complete the diagram of the globe by labelling the lines of latitude shown below.

   a 90°S   ....

   b 66½°S   ....

   c 23½°S   ....

   d 0°   ....

   e 23½°N   ....

   f 66½°N   ....

   g 90°N   ....

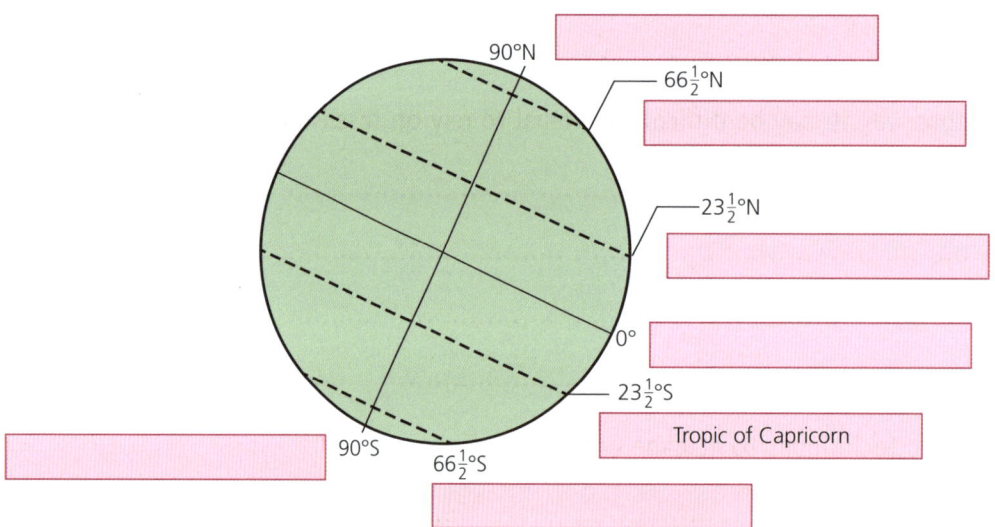

2 Complete the diagram showing bearing and compass direction.

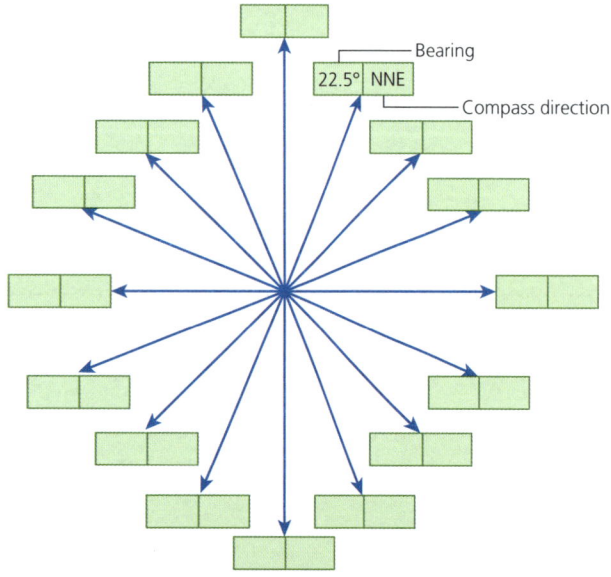

3 Find examples of the following types of maps in the Student Book:

  a Base maps: pages ....

  b Sketch maps: pages ....

  c Isoline maps: pages ....

4 Study the map below.

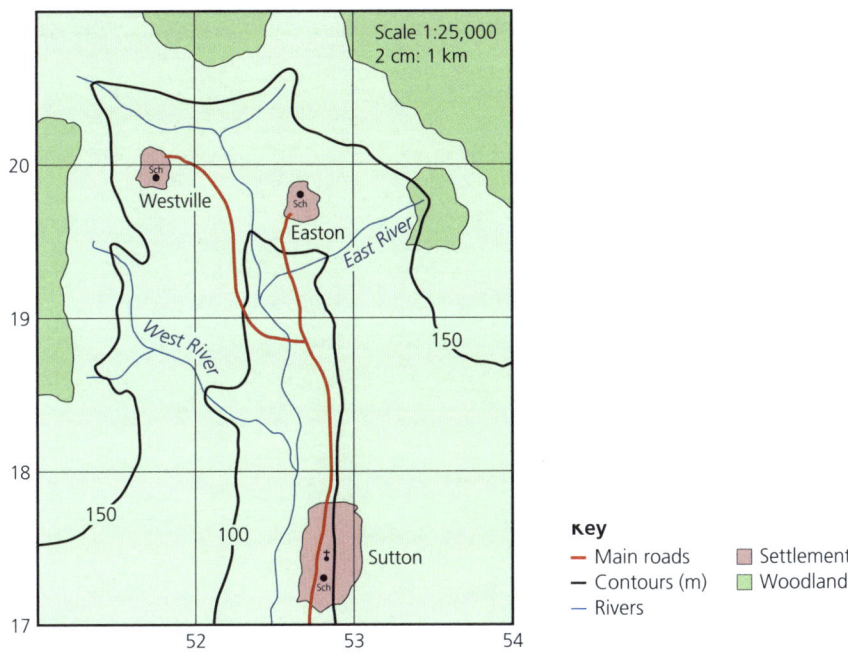

  a State the direction of Easton from Sutton. ....

  b Calculate the distance of Westville from Easton. ....

  c Describe the distribution of woodland shown on the map.

  ....

  ....

  ....

5 Complete the cross section from 510190 to 540190 to show changes in gradient and the position of the main river.

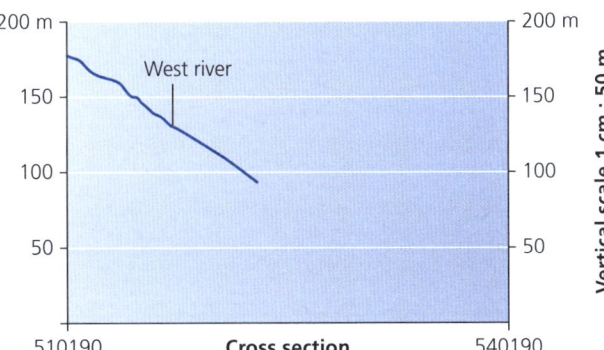

GEOGRAPHICAL SKILLS

6 Draw a sketch map of the area, at a scale of 1:50,000, to show the distribution of woodland and the rivers. The 100 m and 150 m contours have been drawn on for you.

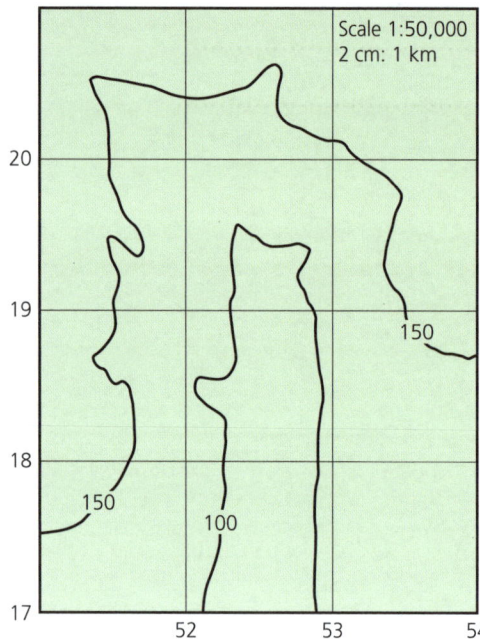

7 The diagram shows global climate change between 1880 and 2020.

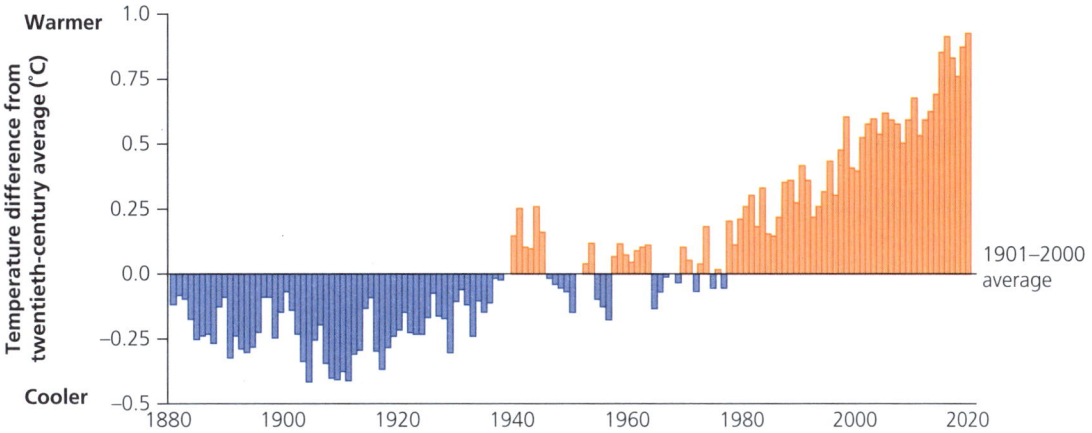

a State what the graph tells us about temperature in (i) 1980 and (ii) 2020.

i ....

ii ....

b By how much did the temperature change between 1880 and 2020?

....

c What does the horizontal line 0.0 show?

....

d Describe the changes in temperature between 1880 and 1940 compared with the average temperature for 1901–2000.

....

....

....

....

....

e What conclusions can you draw about global climate change between 1880 and 2000?

....

....

....

....

GEOGRAPHICAL SKILLS

8 The graphs below are compound line graphs and bar graphs showing renewable energy consumption and generation, 2000–2020.

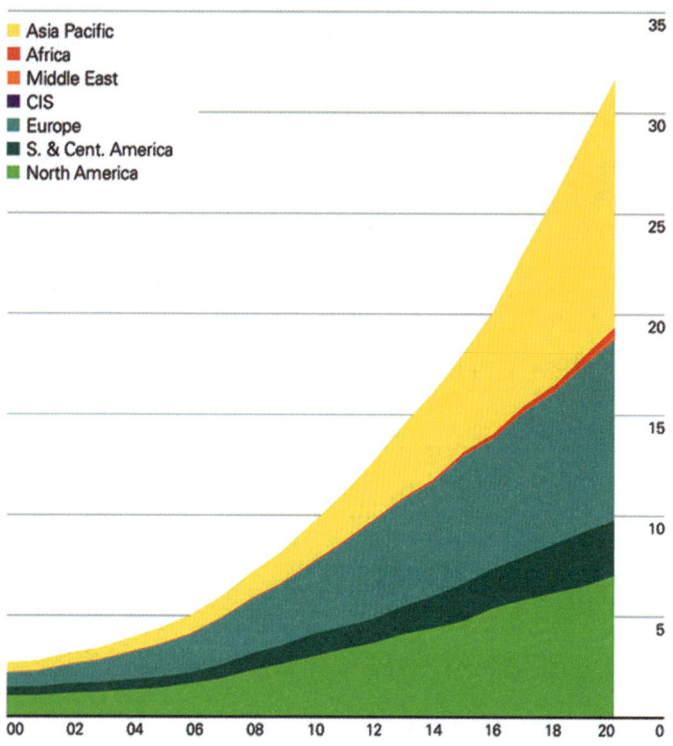

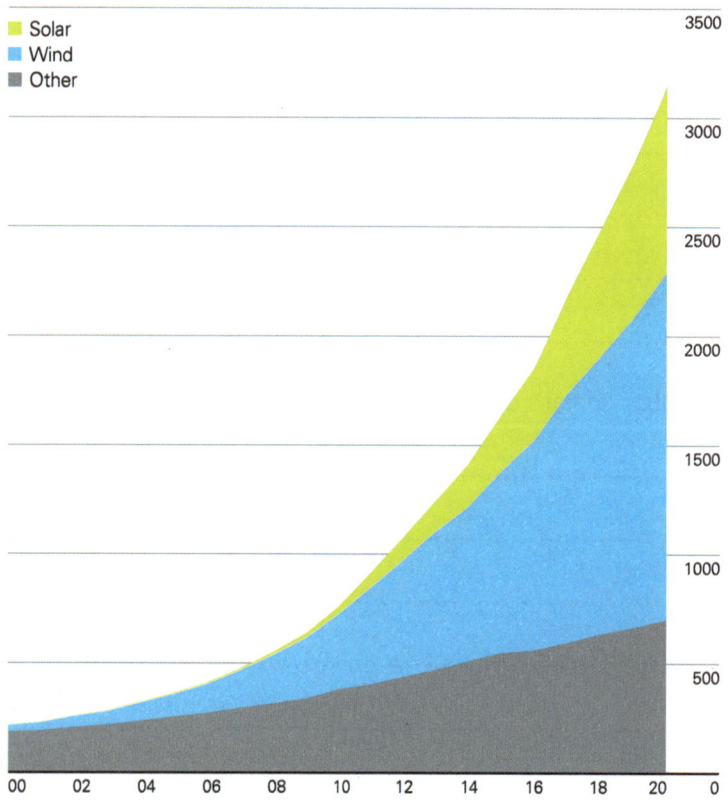

(Source: BP Statistical Review of World Energy 2021, page 54)

Geographical skills

Power generation

Renewable power generation

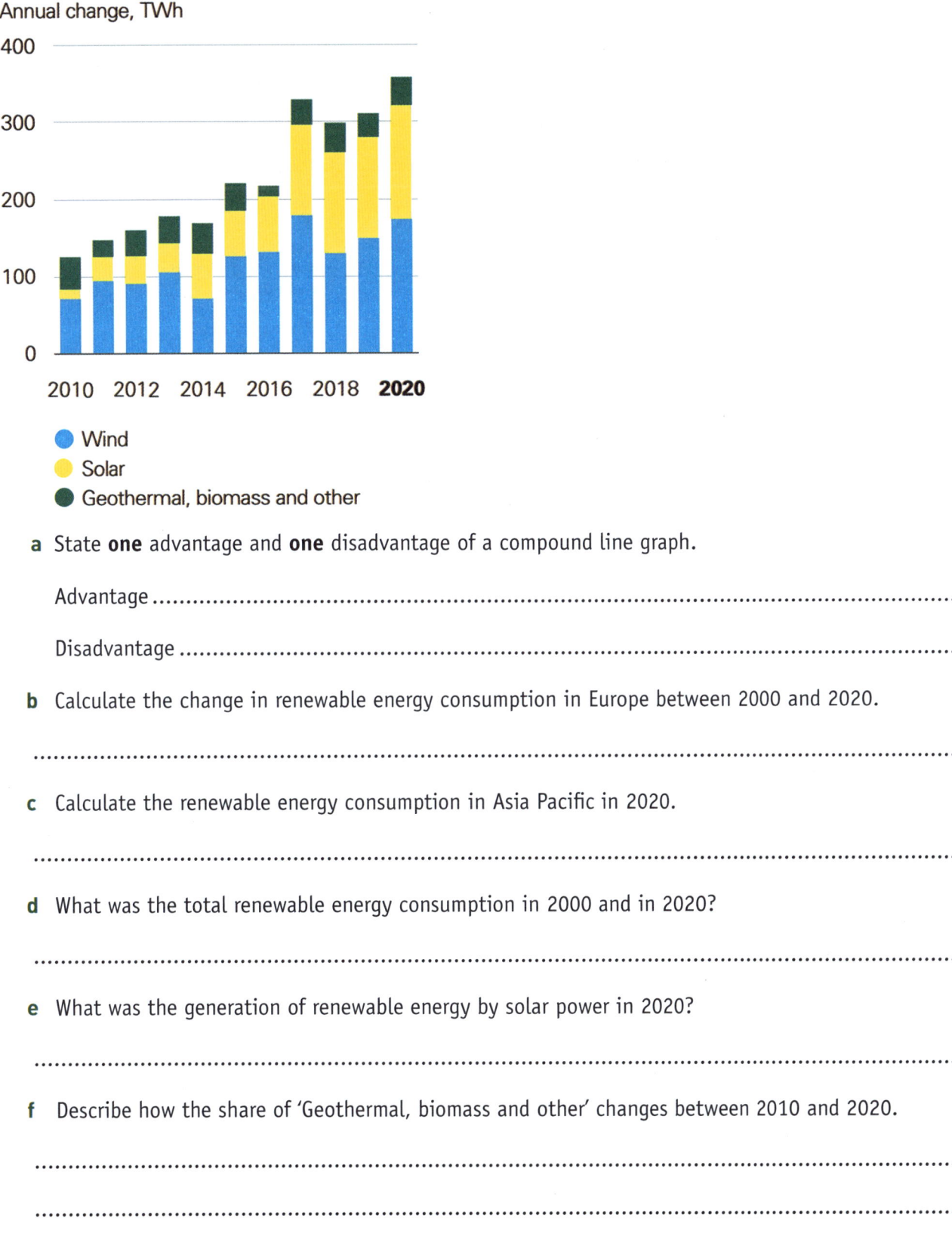

a State **one** advantage and **one** disadvantage of a compound line graph.

Advantage ....

Disadvantage ....

b Calculate the change in renewable energy consumption in Europe between 2000 and 2020.

....

c Calculate the renewable energy consumption in Asia Pacific in 2020.

....

d What was the total renewable energy consumption in 2000 and in 2020?

....

e What was the generation of renewable energy by solar power in 2020?

....

f Describe how the share of 'Geothermal, biomass and other' changes between 2010 and 2020.

....

....

....

GEOGRAPHICAL SKILLS

The diagrams below show some of the key features of population pyramids.

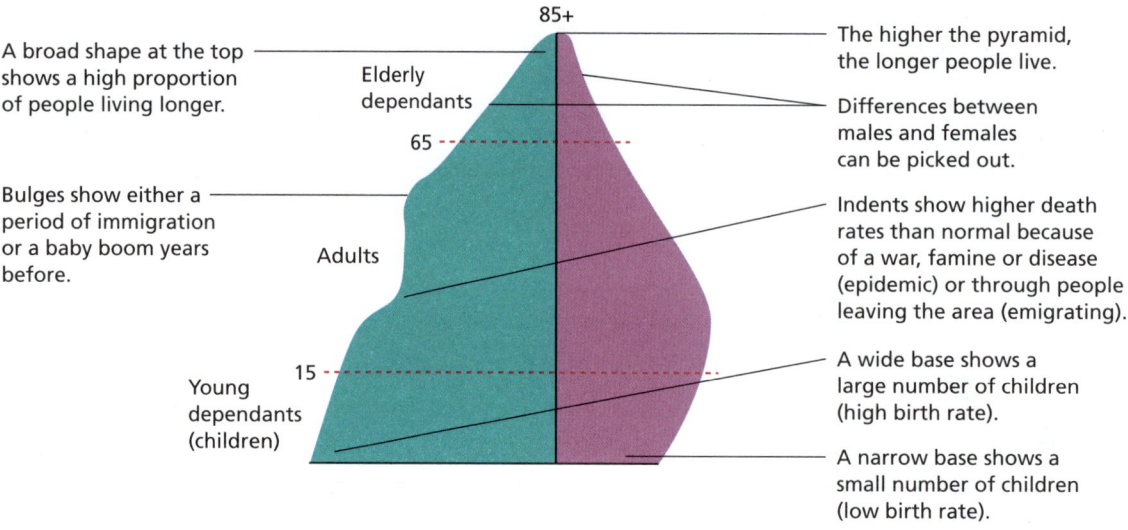

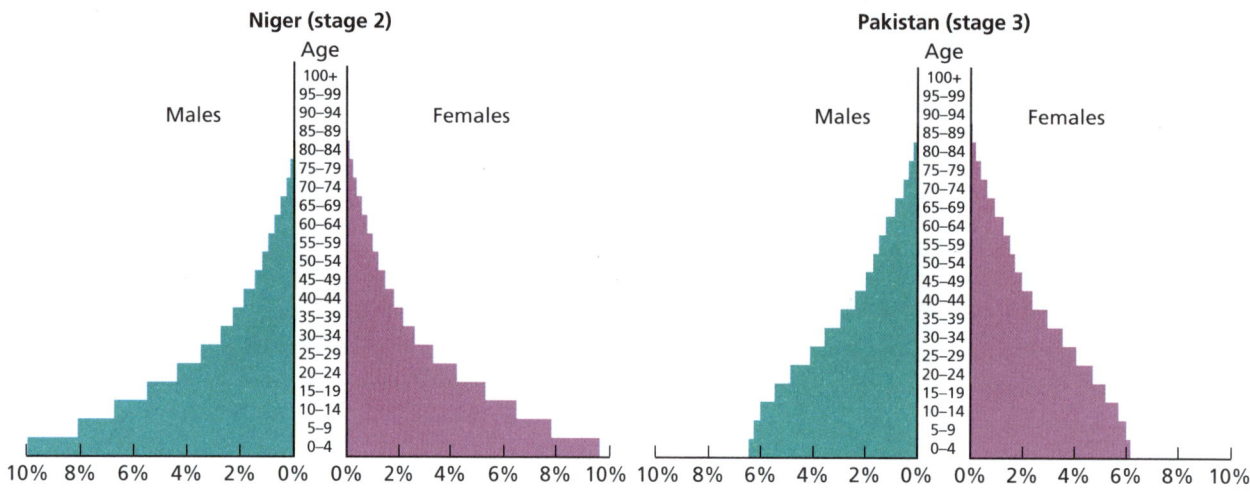

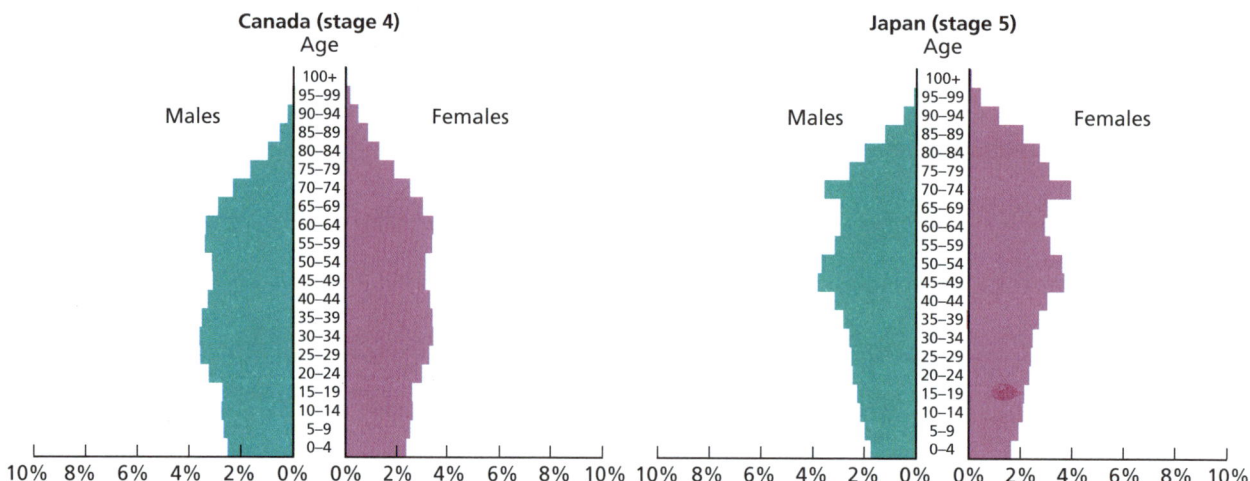

NB: The 'stage' refers to the stage of the demographic transition model that both countries are in.

9 a Identify which of these countries has a youthful population and which one has an ageing/elderly population.

Youthful ....

Ageing ....

Geographical skills

b Compare the percentage of children aged under 14 years in Pakistan and Japan.

....

....

c Compare the percentage of elderly aged over 65 years in Pakistan and Japan.

....

....

d What is the age of the largest cohort (population group) in each of the countries.

....

....

e Identify any gender imbalances in the two countries.

....

....

10 Identify **one** ground level photograph and **one** aerial photograph from the Student Book.

Ground-level: page ....

Aerial: page ....

11 Distinguish between aerial photographs and satellite images.

....

....

....

....

....

....

12 State the meaning of 'GIS'.

....

....

GEOGRAPHICAL SKILLS

13 Explain the benefits and limitations of GIS.

....

....

....

....

....

....

14 Define:

a median ....

b mode ....

c mean ....

d range ....

15 On a copy of Figure 8.4 from page 182 of the Student Book, draw a line of best fit to show the relationship between doctors per 1000 people and GDP per capita.

16 Distinguish between random, systematic and stratified sampling.

....

....

....

....

....

....

17 Explain the value of a pilot study.

....

....

....

....

....

....

18 Identify **three** pieces of equipment used to measure slope angle.

1 ....

2 ....

3 ....

19 Briefly explain how you would measure the cross-section of a river.

....

....

....

....

....

....

....

Reinforce learning and deepen understanding of the key concepts covered in the latest Cambridge IGCSE ™, IGCSE (9-1) and O Level Geography syllabuses (0460/0976/2217) with this updated Workbook. An ideal course companion or homework book for use throughout the course.

» Develop and strengthen skills and knowledge with a wealth of additional exercises that perfectly supplement the updated Fourth Edition Student's Book.

» Build confidence with extra practice for each lesson to ensure that a topic is thoroughly understood before moving on.

» Improve geographical skills such as data interpretation, and diagram and map reading with practical applications and exercises.

» Keep track of students' work with ready-to-go write-in exercises.

» Save time with all answers available FREE to download from: www.hachettelearning.com/answers-and-extras

This text has not been through the endorsement process for the Cambridge Pathway.

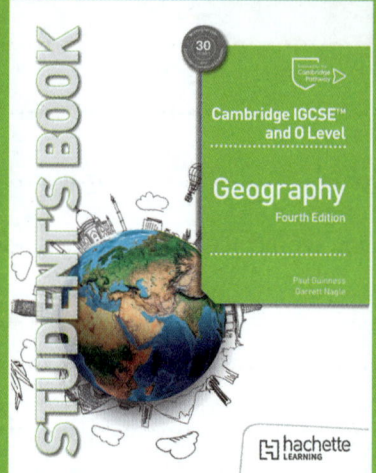

Also available:
Cambridge IGCSE Geography, Fourth Edition
9781036010836

The Student's Book is endorsed for the Cambridge Pathway.

For over 30 years we have been trusted by Cambridge schools around the world to provide quality support for teaching and learning.
For this reason we are an Endorsement Partner of Cambridge International Education and publish endorsed materials for their syllabuses.

Visit us at Hachettelearning.com

ISBN 978-1-0360-1084-3